Riding the Winds of Change

The Chronicles of Jack Hall

JOHN HALL
with
PATRICA VAUGHN

cover design by
Mary L. Weldy

map illustrations by
Russ Matthews

Printed in the United States
by Princeton Academic Press
3175 Princeton Pike
Lawrenceville, NJ 08048
609 / 896-2111

First Edition

Riding the Winds of Change

The Chronicles of Jack Hall

Published in the United States
by ICAN Press Book Publishers, Inc.
616 Third Avenue
Chula Vista, CA 91910
619 / 425-8945

Disclaimer

This document is an original work of the author. It may include reference to information commonly known or freely available to the general public. Any resemblance to other published information is purely coincidental or accidental. The author has in no way attempted to use material not of his/her own origination. ICAN Press disclaims any association with or responsibility for the ideas, opinions or facts as expressed by the author of this book.

Printed in the United States of America
Library of Congress
Cataloging-in-Publication
ISBN 1-881116-51-4

FIRST EDITION

Acknowledgments

I would personally like to thank Partica Vaughn for her editorial consultation and Russ Mathews for doing the internal maps and illustrations.

ICAN Press would like to acknowledge the following contributors for the production and marketing of this work: Dahk Knox, Marian Denson, Michael McKenna, Mary Porta, Sophia Martin, Mike Green, Patricia Godsoe and especially Steven Fellwock.

Table Of Contents

Prologue

John 'Jack' Hall's life spanned most of the twentieth century—a time of world wars, changing national boundaries, growing national independence from colonial rule, and the period in which oil became a world power.

His character and values were shaped by the Industrial Revolution of the late nineteenth century, which drew thousands of immigrants to the United States with its promise of opportunity and prosperity. Jack's own ancestors, of English, Irish and Scottish descent, had been early arrivals in the New World. Though one of his great-great-grandfathers had been postmaster general of Ireland, his grandfather had fought in the American Revolutionary War.

His was no rags-to-riches story, coming as he did from a solidly middle-class family. The values of hard work and enterprise, combined with the optimism of the times, were stamped on him at an early age. Those values guided him through many complex trials and transitions: from midwestern boy to world citizen; from devoted family man to lonely widower and to a happy second marriage; from globe-trotting executive (in an industry which has done much to shape and control world power) to retiree who built a meaningful new life by channeling his vast energies into civic work. He lived to be eighty-five and spent his last five years in a losing battle with Parkinson's disease.

Like the rest of us, Jack led a life shaped by larger forces outside of his control. What makes his life story remarkable is the astuteness with which he assessed these forces and the part he played in them. His history is the history of the twentieth century, with all its exhilaration, terrors and complexities. It is an inside view of the major forces that shaped the world we live in today.

When Jack was born in 1907, the Model T was not yet being mass-produced by Henry Ford. The oil industry was a relatively new kid on the block, dominated by Rockefeller's Standard Oil Company and not yet affected by the Sherman Anti-Trust Act. His career with Shell Oil Company spanned the Great Depression, World War II and the shifts of world power that followed; the infamous oil cartels, the discovery and development of Middle Eastern oil and the formation of OPEC. In his last days, Jack witnessed Desert Storm, assessing it from the vantage point of his firsthand knowledge of the machinations of world politics, military and covert operations, and the interwoven role of the oil industry.

Born November 13, 1907, to Lee and Lelia Starke Hall, he was to be the eldest of three sons. Jack grew up in a comfortable St. Louis suburb in a relatively innocent age.

He began to drive and held his first job at age ten, singing in a church choir for a dollar a week. At twelve he worked for Ford Motor Company, taking the trolley from his Webster Groves home into St. Louis, where he would pick up new cars and drive them to the Webster Groves dealership at a dollar per car.

In high school, Jack worked as a soda jerk and filling station attendant. He excelled at sports and won an athletic scholarship to college, where he joined ROTC. In his senior year at Washington University, he was offered an appointment at West Point.

It was 1931, the depths of the Great Depression. Jack turned down West Point and went to work for Westinghouse. A year later, he accepted a position as an electrical engineer with Shell Oil Company and began his long ride on the winds of change. In 1967, Jack ended his career as a vice president of the company, Chairman of the Board, and Managing Director of Shell Puerto Rico.

Along the way he married Virginia Gibson with whom he shared forty years and the raising of their two children, John Jr. and Kathryn. They had six grandchildren by the time Virginia died of cancer in 1973. Jack served as a respected member of the community in Naples, Florida, where he also lived with his second wife, Millie, who continues to be active in civic concerns there.

In 1987, Jack Hall was struck with Parkinson's disease. Much of the energy he had put into civic matters was then refocused into battling this progressively destructive nervous disease. Always a fighter, Jack utilized the self-discipline and persistence he had honed over a lifetime. With Millie's help, he brought the disease into partial remission through determination and unflagging effort. Until the last few months of his exemplary life, which ended March 22, 1993, Jack stayed active and greeted the world with good cheer.

— Editor

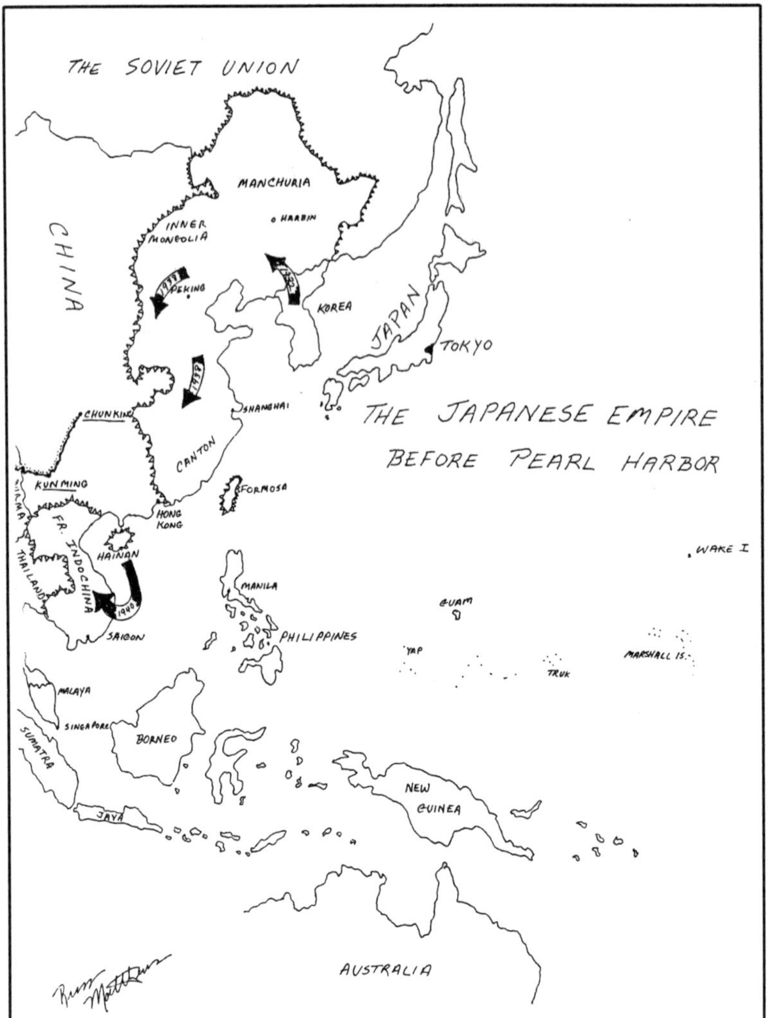
THE SOVIET UNION
MANCHURIA
HARBIN
INNER MONGOLIA
CHINA
PEKING
KOREA
JAPAN
TOKYO
SHANGHAI
CHUNKIN
CANTON
THE JAPANESE EMPIRE BEFORE PEARL HARBOR
KUNMING
BURMA
FORMOSA
HONG KONG
FR. INDOCHINA
HAINAN
THAILAND
1940
WAKE I
MANILA
GUAM
SAIGON
PHILIPPINES
YAP
TRUK
MARSHALL IS.
MALAYA
SINGAPORE
SUMATRA
BORNEO
NEW GUINEA
JAVA
AUSTRALIA

Chapter I

Transformation

I was right behind the crew while boarding the *Hawaii Clipper*. It had been a long wait in the raw October air, and the diplomatic pouch cuffed to my wrist was feeling heavy. My money belt, crammed with identification and clearance papers, cut into tense stomach muscles. San Francisco's grey mist had seeped into my bones. I leaned forward, impatient for a dry seat and a warming drink.

Somebody jostled my arm, renewing the throb of yesterday's inoculations as we shuffled closer to the plane's hatchway. I swallowed a groan and inched forward. At last, I stepped into the waiting plane and fell into the nearest window seat.

Close on my heels, six army officers and ten civilians filed methodically from the dock and onto the rolling plane. We were soon engulfed in a surge of power as the pilot opened to full throttle. Lifting off the water, the *Clipper* began its long slow climb, just as the setting sun set ablaze the Golden Gate Bridge.

We were soon out of sight of land and cruising at 7,500 feet. The cabin began warming up as the steward appeared with a drink cart. Releasing my grip on the armrest, I took a deep breath and looked around at the other passengers.

The six army officers were huddled together and talking guardedly among themselves; there were also a Dutch couple, both reading; a few businessmen and what appeared to be a sprinkling of diplomatic types. I glanced out the window at the endless stretch of water, then got up and strolled toward a couple of friendly-looking guys about ten rows ahead of me.

One of the men was Bill Price, whom I'd met at the San Francisco airport. A tall and powerful man, Price was signalling to the steward with his right hand as his left thumped the armrest, emphasizing some point he was making to the big blond man beside him.

"Hi, Bill," I said. "Mind if I join you?"

"There you are, Jack!" Price boomed, continuing to wave at the steward. "Have a seat and meet Peter Hallberg. Pete's the Swedish consul to Rangoon, just returning to his post after seven months in the States."

"Jack Hall. Lend-Lease, on my way to Hong Kong. Pleased to meet you." I smiled as I shook the big hand Hallberg offered. His grip was firm and dry.

Though I had been coached before leaving the States on the importance of tight security, I didn't think my destination was much of a state secret. After all, my hometown Indiana newspaper had printed a story about the survey I would be making for Uncle Sam in China, and just before takeoff a news photographer from the *San Francisco Chronicle* had shot a picture of me and Price (a pilot), also headed for Hong Kong.

"Pete and I were just making a little wager on the war," Price announced. "He figures the U.S. is going to be drawn into it, and soon. I say we'll keep sending supplies but stay behind the scenes. How do you see it, Jack?"

"Through a knothole, I'm afraid," I said, taking the seat opposite them. "All I know are the rumors and what's in censored news reports. I'm afraid you two have the advantage on me with your diplomatic and military ties. I don't even know enough to make a sharp bet, but I'd like to hear your ideas. What makes you so sure we'll be dragged into it, Hallberg?"

A frown spread across Hallberg's broad pink face. "The facts, my American friend, the plain facts as reported in your newspapers," he answered. "As I see it, your country's freezing of Japanese assets in July amounted to an oil embargo."

This was a new and startling thought for me, but before I could protest he added, "Oh, it was necessary, of course, given the Japanese move into Indochina—nobody expected them to stop there, and your government couldn't allow them to move south into the American and British zones in China. So you stopped selling them scrap metal and froze their assets."

"But Roosevelt hung back from an embargo, didn't he?" I asked, wondering which papers Hallberg had been reading.

"Of course. His is a complex path," he said knowingly. I waited for more.

"When Americans realized your country was selling metal to Japan, they felt they were contributing to the terror bombings in China by supplying this material for Japanese planes. They screamed for economic sanctions to relieve their collective conscience, so Roosevelt put a limited embargo on metals and aviation fuel." I nodded, remembering the heated editorials I had read on this.

"Also, your people don't like Japan's attempts to set up a closed economy in Asia," he went on. "They're angry about the Japanese interference with American rights in the Far East. Roosevelt had to respond to these new public pressures—after all, he's up for reelection—but he's also got to look at the international picture."

So far he was making shrewd sense.

Hallberg lit a Camel, then continued. "Japan depends on the U.S. for oil. If you cut off that supply, they're going to expand into Southeast Asia to get the oil of the Indies and the rubber from Malay, as well as the rice from Indochina."

I nodded again, this time less happily.

Price had finally gotten the steward's attention. As drinks were handed to us we touched glasses, and Price took advantage of the interruption to insert his view. "I'll grant the Japs have their eyes on that stuff, Pete, but we can keep 'em out of there without getting directly involved."

"How?" Hallberg ground out his cigarette, looking as if he hoped it could be true. "The war in Europe has caused just the power vacuum they need to make that move, and it's sure to threaten the whole western position from Hong Kong to Singapore, including the Philippines and possibly even India."

Price downed his first drink in one gulp and signalled for another. "Lend-Lease," he roared, thumping his armrest. "The way I figure it, that $25 million we loaned China in September was only the beginning. Right after that, we also earmarked a hundred P-40's for China, and invited American pilots to go fly them. There are a lot of pilots like me, tired of war-game maneuvers and ready for the real thing. Hell, being loaned to China to fly those babies is a real sweet deal for us, and for the Chinks, too."

Two of the army officers across from us were scowling at Price—probably because he was drowning out their whispered conversation, but possibly because they didn't like his views. Oblivious to them, Price boomed on.

"Our brass knows the Limeys and Dutch can't withstand a Jap attack on Malay or the Indies without our military aid, but aid is all it'll take. Once we get those P-40's flying over the Jap sea lanes, that'll keep 'em in their place!"

Hallberg sipped his drink thoughtfully. "I sincerely hope you're right, but I doubt it. I doubt it so much, in fact, that I have left my wife and daughter in Milwaukee for safekeeping."

"Well, I don't blame you. They'll be a helluva lot more comfortable there, war or peace." Price laughed loudly, then called for a third drink.

I was still quietly nursing my first drink, wondering for the hundredth time what I'd gotten myself into. If Hallberg was right, the world was heading for a two-ocean war. It was a war the United States wasn't ready for, and one for which the British fleet was inadequate. Roosevelt knew that if Britain were defeated, the United States would be living at the end of a Nazi gun barrel. And Britain couldn't make it without supplies from the Far East.

The Japanese had already forced France to close the railroad from Hanoi into China and ordered Britain to close the Hong Kong frontier and the Burma Road, China's last channel to the West.

We had all grown silent. The steady drone of the *Clipper's* four engines covered the whispered exchanges between the army men a few seats ahead, and everyone else was either sleeping or reading. I had a headache.

I thought back to the Japanese teaming up with the Axis only a year ago. That had galvanized the United States into sending aid to Britain and China. Lend-Lease had made this aid an investment, backed by the release of American pilots from regular military service for mercenary service to China.

Maybe Price was right. Maybe these American planes and pilots could protect the Burma Road and keep us out of the war. But on the other hand we had also activated the draft, and last month the Japanese had sealed the China coast.

Only one thing seemed certain. No matter how well the Burma Road was protected, something had to be done about the slow and inefficient movement along it. That was being handled through Lend-Lease too, with loans of civilian experts like myself. It had now become our job to see that the needed fuel flowed along the Road.

I stared out the window at the darkness, feeling disembodied. It was October 27, 1941. As we streaked across the Pacific, I was being transformed from a peacetime domestic pipeline superintendent into being a designer of a vital lifeline that would determine the wartime allegiance of China and the critical fate of Britain and the United States.

Chapter II

By the Beautiful Sea

The steward called us at 6:30 for an early breakfast and our first sight of the Hawaiian Islands. Passing Koko Head, Diamond Head, and Honolulu Harbor, we circled Pearl Harbor for an 8:00 A.M. landing. I had slept off my headache and was thrilled to see all our ships in the harbor. After the dark picture painted by Hallberg last night, it comforted me to see our navy all decked out in war paint and ready to fight.

As we taxied to the dock, I could see Hickam Field, alive with planes of all types. As I crossed the ramp to the Pan Am office, I was looking up at the tropical morning sky. It was dotted with patrol and fighter planes. A lovely Hawaiian girl stopped me and hung a lei around my neck, and I remembered where I was.

Several Shell people met me at the airport. One of them was Jack Montgomery, Shell's sales manager for Hawaii. No doubt the New York and San Francisco offices had notified them of my arrival. I was in for some posh treatment.

Working for an international company like Shell meant sudden travel assignments that were hard on family life but fairly exciting. I got to see a lot of places I would not have visited otherwise and usually enjoyed full company courtesies.

Montgomery, having heard about the bad engine our *Clipper* had developed since leaving San Francisco, offered to show me around during the repair layover. We made plans for the afternoon as he drove me to my hotel.

An hour later from my hotel window, I saw the clear blue water and velvet sand of Waikiki Beach. Diamond Head faced me. The view was glorious; puffs of clouds herded by, trade winds blew across mountain peaks, and rich vegetation swayed in the breeze. I'd never seen greener greens or bluer blues. In no time, I was pawing through my luggage for swimming trunks.

Working my way through layers of Florida shirts and Canadian sweaters, heavy denims and boots, I smiled as I remembered Virginia and the kids scampering around to help me pack. Seventy-seven pounds of luggage seemed like a lot for a two-month trip, but I'd been told to take clothes to cover everything from tropical to freezing weather, and also rough clothes for camping in the mountains along the Burma Road.

I'd never have managed getting under Pan Am's weight limits without the help of Virginia and the kids. The packing had been exciting and mysterious for Kathryn and John. At seven and eight, they couldn't comprehend how long I'd be gone. Virginia's cheerfulness had been an act of generosity. She hadn't complained either, about my being gone for our tenth anniversary, the first we'd missed since living together. That wasn't like my outspoken darling. But then, neither of us had been behaving normally. In spite of my offhandedness, she knew the hazards of this assignment.

Hallberg had been on target about the effects of our ending the trade treaty with Japan. After four years of war with China and with no end in sight, the Japanese were chafing at the economic restrictions we'd imposed.

Though news reports back home weren't painting the situation as critical, Virginia's worries were valid. British and Dutch possessions in the Far East were blocking Japanese development of Thailand and Indochina, and the Japanese didn't seem to be in a mood to accept that.

The evening before I left (was it only last Thursday?) Virginia and I had talked about these developments. Realizing we couldn't second-guess the Japanese timing, we accepted the unknown with wry banter, then clung to each other far into the night.

I remembered once more that I was now in Hawaii. I pulled on my swimsuit and was soon plunging into the clear blue water and swimming with all my strength, trying to suppress the heartache of these thoughts of Virginia.

After lunch, Jack Montgomery drove me around Oahu and over the Pali Pass. We passed through sugar and pineapple plantations, interspersed with emergency airfields filled to capacity with fighter planes. Taking off from carrier decks, planes roared overhead all day doing their maneuvers; they would then return to their ships or land in these emergency fields. From the Pass I saw aircraft carriers and battleships come and go in great numbers. I was convinced our Pacific Fleet was close by, expecting trouble.

Back at my hotel after another swim, I felt too warm and lazy to think about trouble. After a few Mai-Tais in the bar, my mind was capable of only blurry wishes that Virginia and the kids were here enjoying this paradise with me.

♠ ♠ ♠

"Ah, Jack, there you are," Hallberg said, as he stopped me in the lobby that evening. "I'd like you to meet Maj. Gordon Enders and Lt. Edgar Hampton. Join us for dinner?"

Dinner was excellent. Flanders and Hampton, while pleasant though closed-mouthed about military matters, mentioned only that they were headed for new assignments in the Philippines. The rest of their conversation casually griped about the tedium of layovers and lonesomeness. Under close instructions to keep our mouths shut, we chatted about sports (it was football season) and kept our talk light and innocent. None of us were trying to make close friends on this trip, but we were all glad to have pleasant company to help pass the time.

After eleven days of sunburns, shopping, touring and eating too much, our *Clipper* had been repaired and we were on our way to Hong Kong. Then followed four days of constant travel, with stops at Midway, Wake and Guam. Because of all the military activities, our personal movements were restricted at each stop. There wasn't much to do except eat and sleep. By the time we arrived in Manila my sunburn was gone and I was carrying an extra ten pounds. Six feet tall, I looked a little rounded at 190 pounds. I figured the padding was insurance against lean times ahead.

Chapter III

Into Darkness

In Hawaii, it had been hard to take prospects of war seriously. Even military personnel were tanned and relaxed. The farther west we moved, the darker things looked.

On our way to Guam we passed the eastbound flight, said to be carrying Kurusu, the special Japanese envoy on his way to Washington for talks of settlement of the Pacific status of Japan. We expected unending talks with nothing accomplished, but some of the army men thought Roosevelt was going to take a firm stand. They spoke of a showdown by the first of the year. I was hoping things wouldn't escalate for at least a couple of months, time to get this job done and be home before war closed the Pacific route to civilians.

At Guam we were checked in by the marines. They seemed ready for trouble, but there seemed too few of them to defend the island.

Only a short distance from Japanese territory, Guam was our best bet to become a first-class, long-range bomber base—if only we had enough time to get our permanent defenses in place there.

In Manila, we landed at Cavite, the naval base for our Far Eastern fleet. The *Clipper* was scheduled to continue on to Singapore in the morning. Price and I said good-by to the crew that had flown us 2,400 miles. We were going directly to Hong Kong on another plane.

As we cleared immigration, reporters milled around looking for war news. Apparently none of the army men were important enough to impress them and they soon left. We drove to the hotel through intense heat and humidity. The mix of naval bases and local villages looked strange. Huts and unpainted wooden shacks contrasted sharply with white-stone government buildings that would have fit in Washington, D.C.

National elections were going on, which meant that our hotel was crowded. Most of the usual entertainments were shutdown. I shared a room with Hallberg, and we spent the evening at the Jai-Alai house, which was going full blast.

Price met me at the airport the next morning and introduced me to our pilot and to Emil Scott, another pilot with China National Airways, who would be flying along with us. Scott, on leave from CNAC, had been visiting his wife in Manila.

As we took off, we passed Clark Field and saw long rows of "Flying Fortress" planes, their aluminum skins shining in the morning sun.

"They'd make swell targets in a surprise attack," Scott said. "I can't understand why we don't camouflage and scatter our planes, like the British and Dutch do."

Scott had flown bombers for our air force before resigning to fly for China. He believed we would soon need all the planes we could get. He had heard from the Chinese that there was a lot of action near the Kowloon border. That was too close to Hong Kong for my comfort. I escaped into a nap, leaving Price and Scott arguing about probable "encounter" dates.

Lifting off from Macao, where we'd made a mail drop, I woke up just in time to see the nose of our plane veer up sharply. The engine pulled with a sickening whine that brought my stomach up to my throat as we skimmed over the masts of a native junk, just missing it. We were over Hong Kong harbor before my breathing returned to normal.

"Will you look at that," Scott said to Price, pointing to the blanket of submarine nets and mines in the harbor. "I can't figure how they get even friendly ships into this harbor. But those big guns on both sides of the entrance . . . why in hell is every plane that flies over Hong Kong allowed to see these defenses?"

Scott and Price continued their military assessments as we taxied to the dock of Kai Tak Airport and went through the usual passport routines. An airport car took us to the ferry, which carried us to Victoria Island and the city of Hong Kong.

There was a message waiting for me at the hotel. It was a dinner invitation from Bousfield, of the London office of Shell Transport and Trading Company. I'd met him once in the New York office during one of his frequent trips to the States to check out new technical developments. I'd been expecting his call but was surprised by the invitation. Bousfield was a big man in the Shell hierarchy and travelled all over the Far East. All the local employees reported to him.

As soon as I got to my room I called Shell House. Bousfield was out, so I left word and headed for the Bank of China, just down the street. In the bank, I was ushered into the office of two English-speaking officials. With much smiling and bowing, they expressed great interest in my trip and said they would make all necessary arrangements with China Defense Supply for my expenses.

With these matters taken care of, they barked a command at the page who had been standing quietly near the door. He slipped out and returned a moment later with another man. I was introduced to Fisher Yin Chen Yu, an official of the bank who was assigned as my Hong Kong host.

In impeccable English, Yu said, "I am honored to be of service to you, Mr. Hall. May I suggest we go immediately to the police department?" He bowed deeply.

Hong Kong is British territory and I am considered an alien here. Yu's concern that I comply with local registration laws was understandable, and I was relieved to have a guide and helper. After a number of good-by bows to the other bank officers, we set out to fill in the necessary forms and have my passport stamped.

The police department was crowded with long snaking lines, but after a few words from Yu to the officer at the door, we were escorted to the captain's office and were soon back on the street, on our way to the China National Airways office.

I was feeling like a seasoned traveller by now, comfortable with the unfamiliar sights, smells and customs. But the CNAC flight procedures threw me.

"What do you mean they don't know when the plane leaves for Chungking?" I asked Yu.

"Ah, yes, they are very mysterious about these things," Yu smiled. "It is an unfortunate necessity, sir."

"Necess . . . look, Yu, I have to set up a lot of appointments here in Hong Kong. How can I schedule them if I don't know when I'm leaving?" I kept my voice under control but couldn't suppress my annoyance.

Yu looked at me with total serenity. "You see," he began, "we can only fly when no Japanese planes are along the route. They consider CNAC planes real prizes, and would rather shoot one down than destroy a humble village. So it is necessary to wait for signals from the villages along the way, informing us that the route is clear, before we can consider embarking. Also, of course, we confound Japanese informants with irregular and unscheduled flights."

This was so reasonable that I couldn't argue. The thought of being shot at was not appealing. But I pressed Yu to find out at least what day was likely for our departure to Chiang Kai-shek's wartime capital. I stood grinding my teeth while Yu carried on an interminable conversation in Chinese which seemed to require far too many bows. Finally, he turned to me and said, "Maybe Friday."

Out on the street again, I feel less seasoned. We are crowded and pushed by diminutive local people of all ages, speaking words I don't understand. The packages they carry smell of fish and other unpleasant things I can't even identify. My blond head is the only one in the crowd.

Shouting over the din of the streets, Yu asks what clothing I have brought and advises me to buy some tropical suits and shorts for Burma. He guides me to the relative quiet of a tailor's shop, where

I am measured and promised my new clothes by Friday morning. When it is time to pay, I am pleased to find out how low the prices are, so just before leaving I order a second suit to be picked up on my way back from Burma.

I blink in the brightness of the street, suddenly tired and hungry. Yu offers to teach me the use of chopsticks. I have had enough novelty for one day but I cannot refuse his smile. I barely manage, despite Yu's excellent instructions, but am able to shovel in enough food to get full.

Back in my hotel room, I fall into a bamboo armchair, feeling isolated, and worried about Friday's flight. The war that seemed so far away from Indiana is here, and I am in the midst of it!

I glance at a Chinese newspaper on my bedside table. It is meaningless to me except for the picture of T.V. Soong, Director of China Defense Supply, on the front page. He is shaking hands with Franklin Delano Roosevelt, and behind them stand Warren Lee Pearson of the Export Import Bank and Henry Morganthau of the U.S. Treasury.

T.V. Soong is the brother-in-law of Chiang Kai-shek, whom the Japanese had chased to Chungking in 1938. He is Chiang's emissary to the States, where he has capitalized on his Christianity and Western education to ingratiate himself with Roosevelt and his cabinet. Soong is reported to be the second richest man in China and heavily implicated in China's heroin trade.

Now Soong was the recipient of $25 million American dollars, and it was rumored that Roosevelt was only awaiting his reelection to ask Congress for more. The rumor was that the Delano family had ties with the Shanghai opium trade; and that had predisposed Roosevelt to give Soong more, and now Soong was my boss. I was just beginning to feel a cold sweat break out, when the phone rang.

"Welcome to Hong Kong, Hall. This is Bousfield. Sorry I missed your call this morning. Can you join me and a few of the Shell boys for dinner tomorrow, at the guest house?"

I managed to accept graciously; without indicating the slobbering gratitude I felt at the prospect of being with people who wouldn't seem so foreign. What had begun back home as an exciting challenge was now turning into a dark maze. I couldn't wait to talk to Bousfield.

Chapter IV

The Assignment

I felt calmer after a good night's sleep. I spent the morning reviewing my assignment, and during the afternoon I walked the business district near my hotel, thinking about what I'd been sent to do. I reflected on what I'd been told so far about my assignment, and how all this had put me where I was today.

Early in 1941, British intelligence had informed the U.S. government that British Petroleum was unable to deliver oil fast enough to meet British military needs. Without this oil, Britain would be unable to hold off the Nazis. Something had to be done. British secret service agents had also contacted Edward Mayne, executive director of Organizations with British Shell, asking for help.

Mayne met with U.S. Shell, explaining that BP couldn't guarantee needed oil deliveries without improvements and protection on the Burma Road. Both air forces had told the U.S. government the same thing, so the two governments had begun a cooperative program to get the oil flowing.

The problem was with the Burma Oil Company pipeline. This pipeline carried crude oil from Singapore to a refinery in Rangoon, where it was then transshipped north on the Burmese railroad. In drums fastened to flatcars, the refined oil went north to Lashio where it was loaded onto trucks to be transported over the Burma Road. At Lashio, tanks were set aside to transfer oil to the Burma pipeline, which was the only way to get oil up the mountains and over to China.

Our government had been involved in this for some time. The U.S. Corps of Engineers had built the road and pipeline through northern India, toward China. At the China border, the Chinese had taken over and continued the building through their country. But their government inefficiencies and constant strafing by Japanese planes had caused a bottleneck, and oil was piling up in Lashio.

At Mayne's suggestion, our government asked each of the oil companies in the States to appoint one of their executives as a "dollar-a-year man," (through Lend-Lease) to study their own company's existing pipeline facilities. The hope was that American technology had already produced facilities that could be used on the Burma Road pipeline; these only needed to be identified and reproduced. Like everything else connected with the war, this needed to be "done yesterday"—there was no time to research and develop anything new.

So British Shell looked at a pipeline design we had been slowly developing for stateside use in Indiana. It looked good. We reported to our government that we could build a pipeline superior to the rubber hose that had been effective for Hitler, following the blitzkrieg, and that it would be far safer.

The government gave Shell's plan the go-ahead, and a proposal was made for a 1,000-mile pipeline that could stretch over some of the roughest country in the world, a roller-coaster terrain ranging in altitude from 2,500 to 8,500 feet.

The proposal had been handled by a group of Shell's technical experts. We set up in Shell's offices in New York and Washington. I never knew exactly how many of us were there. The full and part-time staff varied because we never knew, even with our top priorities, when suppliers' shipments would arrive, and these experts were reassigned during the waits. That period of shuffling back and forth between New York and Indiana had been hectic, but it was little preparation for this trip to the actual site. I still wasn't sure how I'd been picked for the on-site assignment.

The oil companies had all pitched in to meet the Allies' needs. Shell had delivered a pipeline plan, and the American government had provided the Flying Fortresses to protect the remote area. Britain continued to hold off the Nazis, its citizens putting up with ever-increasing hardships. Now, here I was, U.S. Shell's Lend-Lease man, charged with getting Chiang Kai-shek's approval of this plan and getting it rolling.

That was all I'd been told before leaving home, and it was beginning to look insufficient. Maybe Bousfield could tell me more. I looked at my watch. Our appointment was only an hour away.

As the afternoon sun weakened, the busy Hong Kong streets began to lose my interest. I caught a cab back to the hotel, showered, dressed, and was ready when the desk called to say a car had arrived for me.

Chapter V

The Big Picture

My Chinese driver spoke little English, leaving me free to sightsee along the drive.

Hemmed in by cars and taxis, we inched through crowded streets as bicycles and rickshaws crossed our path again and again. Although it seemed strange to see human beings pulling these two-wheeled vehicles, especially humans bowed under large coolie hats, it was stranger still to see Asian men in western business suits riding bicycles.

The houses grew larger as we drove uphill through the residential sections. In Hong Kong, those who could afford to lived on the ridges, where there were panoramic views of the harbor and the airport across the bay. We were in rolling country, travelling from peak to ascending peak. I caught views of the many large planes at the airport. The harbor, crowded with junks, twinkled with lantern lights. The night was clear and brisk; car headlights sparkled on the ridges. Hong Kong looked like a permanent Christmas display.

We topped the highest ridge and stopped. We had reached "the big house," (Shell's guest house). It was no architectural jewel but it had an expensive, well-built and well-maintained air about it.

My driver accompanied me to the ornately-carved front door and spoke a few words in Chinese to the houseboy who had answered our ring. Guided over thick oriental carpets into a large, comfortable drawing room, I could hear the muted sounds of the household staff beyond.

In this room of silks and brocades, it was easy to imagine the kings and sheiks Shell entertained here, potentates and their emissaries from all over the Far East, where oil was king. The sumptuous surroundings catered to their comforts and reminded them that oil was the source of their wealth.

The door swung open and a vigorous, large-boned man came toward me. "Welcome, Hall," Bousfield said, in his strong English accent. "Good to see you again. Delighted you could make it! I'm sure you have a question or two about this territory, so I've invited Mowren and Sheehan from the Hong Kong office to join us for dinner. But first, may I offer you a drink?"

The lamp over the bar highlighted the touches of grey in Bousfield's brown hair, as he poured two Glenlivets. I realized how young he was to be so well-known and respected worldwide in the oil industry. As area marketing manager for a large part of the Far East, he spent a fair amount of time in Hong Kong. Bousfield worked out of London Shell, responsible for advising other Shell companies on commercial activities, shipping and transportation throughout the Far East. He seemed very much at ease. His manner was pleasant, but he wasted no time.

"How much were you told about the political situation here?" he asked, handing me my drink.

"Whose politics?" I'd found out just enough by now to realize how complex Bousfield's question was.

"Capital question, Hall, capital. I'm going to let Mowren explain the China-Japan picture to you, and Sheehan can discuss the American-English cabal. He's current on Britain's military alliance with your country and knows how the United States is involving itself in China. I suggest that you and I discuss how the actions of these various powers are affecting something closer to home, the grand old oil industry."

"Right up my alley," I said, settling back to listen.

Bousfield sat in an armchair across from me. "First," he said, sipping his Scotch, "you must realize that the oil industry is already at war, indirectly. This damned war has seriously undercut our autonomy and we're determined to protect our interests.

"Let me illustrate the changes that have taken place in the last two years. A few months before Hitler started this business in 1939, oil men negotiated the entire Arabian oilfield concession. The entire 440,000 square-mile concession, Hall—as big as your Texas, Louisiana, Oklahoma and New Mexico combined—and we did it without government interference!

"Ah, that was a time . . . remember how Philby put it? 'American technologists descending from the skies on their flying carpets with strange devices for probing the bowels of the earth in search of liquid muck for which the world clamors to keep its insatiable machines alive.' Yes. Unforgettable! Oil had grown into a world power by then. Until this war, nobody told oil what to do!

"Opening the Arabian oil flow was an historic diplomatic occasion, without the bother of diplomats. All negotiations took place between a country and a company. That's the way oil does business, like a private government. Oh, there were annoyances like the U.S.'s antitrust meddling, but nothing we couldn't strong-arm our way through. Governments of industrial nations needed us and they knew it. But with this European war, we find ourselves having to cozy up to some of these friendly governments. We don't like it, but we'll do what it takes to restore our position."

I nodded, knowing only too well the force that lay behind this typical English understatement.

"That is why the oil companies are so generously underwriting the Allied cause," he continued, smiling and waving expansively, "lend-leasing key company men, like yourself, and donating equipment and supplies where needed."

Bousfield stood and began pacing the room. His urbane manner suddenly took on a predatory cast as his smile faded.

"Oil's primary interest is restoring Europe to its position as the center of world trade. All the major companies are anxious to get back into Europe where, except for good old England, Hitler has seized their very headquarters. We believe that what is good for oil is good for the world, and the American and British governments seem to agree.

"Your government sends trained men, planes, ships, and other military supplies, but these aren't of much use without the fuel to fly those planes and launch those ships. That's where oil comes in. Oil interests are helping through the back door, supplying England with loaned equipment and bulk oil, which we send in through Scotland in order to protect our big tankers from Nazi planes and submarines."

Having worked on England's pipeline before being assigned to the Burma project, I already knew some of this. Not enough, however, to interrupt.

Bousfield continued talking as he refilled our glasses. "Actually, Britain has already been whipped by the Nazis but she doesn't have the sense to quit. The oil industry doesn't wish her to quit, and we are backing our wishes with everything we've got! This damnable war has brought about remarkable cooperation in an historically cutthroat business. Do you know why, Hall?" He leaned toward me, his warm eyes turning hawkish.

"Because war is run on oil," I answered quietly, "and whoever gets the oil wins the war."

"Quite so, quite so. The Allies, of course, are deeply indebted to us and they know it! Our fate is inextricably bound to their cause, and we'll call in that debt after the war."

I didn't move a muscle during his pause, while we both thought about the implications of that.

"Shell, as you know, is one of the world's largest oil suppliers, yet our war efforts have put a terrible strain on our resources. We send England mostly Texas gas, but some fuel comes from the East Indies, shipped through Shell Trading & Transport or on rental tankers. And now those inscrutable Japanese are doing to the Far East oil supply what the Germans have done to organizational headquarters in Europe . . . abominable people those, abominable."

"They have either captured, bought, or leased all the oil-producing areas in the Far East and they're picking off our refineries and storage terminals. England is desperate and British Petroleum is outraged . . . their colonial attitude, I fear. Oh, the Japs continue to allow BP and the rest of us to operate our Far East facilities, but I'm worried, you see."

He stared at his drink for a long minute. When he looked up, the predator turned victim; his voice became low and strained.

"I remember how easily the Germans overpowered the Allied forces and took over the low countries in Europe. This gave them a clear passage to Sweden, Norway, Finland—then France and ultimately, they believed, to England."

"I am concerned that the Japanese plan to do the same to China. Did you know, old boy, that twenty percent of Shell's wartime worldwide product is being sent to China? No? Well, it is. The problem, however, is that this oil is not reaching China. We will not allow an island kingdom of buck-toothed heathens to—but I digress. I'll let Mowren tell you about that."

"My point is that oil needs friendly governments in order to prosper. That's why oil backs the Allies. That's why, as soon as we get rid of these uncivilized Germans and Japanese, Europe will resume as the center of world trade and oil will again go its merry way. You must keep this in mind as you attempt to put that pipeline through."

"What was that you said? 'War is run on oil and whoever gets the oil wins.' Yes. Well put. I must remember to share that with—Ah, here they are now! Come in, gentlemen, and meet Jack Hall, a wise and succinct man."

I had time only to notice that Mowren and Sheehan were both pale and nondescript before my attention shifted to two delicate Eurasian women who followed them into the drawing room. They seemed to float into the room in their brilliant floor-length silk sheaths, slit disturbingly to their hips. I didn't know enough about local fashions to be able to interpret status from clothing, and I wondered what they were doing here. This didn't look like the usual Shell protocol.

I managed to tear my eyes away from those flashing thighs long enough to shake hands with my colleagues from the British Isles. The women settled on the floor next to the seated Mowrey and Sheehan. I strolled the room for a few minutes, trying to regain my composure. Then I chose a chair off to the side, where the women were not visible. This helped a little, but their perfume had kidnapped my concentration.

Catching only every other word through another round of drinks, I learned that all three Shell men were pessimistic about solving the supply problems along the Burma Road. During dinner, I found out they were also matter-of-fact about the United States' eventual entry into the war.

The next thing I remembered was getting up from the dinner table and hearing Bousfield say, "You've had a long day. Why not stay here tonight? Chan will see to your needs. Good night, Hall."

Chan guided me to a room on the second floor and bid me good night at the door. In mid-yawn I realized there was someone in the room.

"Good evening, sir. My name is Lo-yi. I am your personal hostess during your stay." She bent to turn down the covers on the large plump bed. Then, pointing to the steaming towels on the bedside table, Lo-yi smiled up at me. "I am expert at massage," she said softly. "May I so serve you? It is my job to do whatever is required to bring you sweet dreams."

She was beautiful. I was starting to like this assignment.

Chapter VI

The Project

Over the next three days, I met several times with Mowren, Sheehan and the other top local Shell people.

I learned that FDR was pushing through a $1 million loan to China, half from the Import Export Bank, to pay for supplies Soong had bought in the States. I liked that, since all the supplies I needed would be financed by Soong's China Defense Supply. The other half, from the U.S. Treasury, was intended to help stabilize China's currency (wildly inflated by the uncontrolled printing of paper money). It was no secret that China's money had become nearly worthless since Japanese conquests had disrupted Chiang's most reliable source of revenue, the opium and heroin trade.

Mowren felt that this loan amounted to an outright subsidy, and served to replace income from narcotics exports and become Chiang's main source of foreign nourishment following the recent American crack down on drug imports.

This was not the only form of American aid. Sheehan reported that the Burma campaign was also in full swing. American Volunteer Group (AVG) personnel were flooding into China. They had established a low-level flying route and opened the pipeline at Myitkiina. The Chinese armies were receiving tons of U.S. equipment which was of little help—the Chinese commanders were selling much of these Lend-Lease supplies on the black market and to the Japanese. Sheehan was not optimistic.

He explained that the core of American policy was to keep Japan from provoking conflict due to her southward expansion, at least until April of 1942, in order to give us time to equip and expand our army, navy, and air support. The strategy of the Joint Board of the Armed Forces was to defeat Germany while maintaining strong defenses against Japan through the Chinese. Nobody expected much from the Chinese Nationalist military, which was FDR's primary political front for the real defenses carried out by American personnel with American war tools. The myth was maintained that the United States was not at war but merely lending a hand to China.

Asia was definitely of secondary concern with the policy-makers, who saw American security as dependent on the British fleet. British survival was a vital American interest in a way that China was not. The Hong Kong Shell view, however, was that too little was being done too late in the Far East to keep the United States out of the war.

Oil, of course, was the hinge on which everything turned. Only so much was allocated to China, and grudgingly at that. If the Japanese were to be contained, the current level of supply would have to be stretched through more efficient use. It was currently taking a ton of gas to deliver a ton of cargo to China. The Shell people here knew we would have to do better than this, but didn't think we could succeed in laying 1,000 miles of pipe between Lashio and Kunming to complete the Burma pipeline.

They considered the whole project to be a wild leap of engineering. The pipe was to stretch 1,800 miles from Calcutta to Kunming, some of it across mountains and canyons. Those who had travelled this route were skeptical that a steel pipeline could be carried across the many riverbeds on the existing wooden trestles and suspension bridges of the Burma Road.

I halfway shared their view. The Burma pipeline was to be the world's longest pipeline, crossing over some of the worst territory on earth. It had already been dubbed "Pipe Dream" by its builders, and was known locally as "Pick's Pike" after General Lewis Pick, commander of the Road and pipeline. Like the Road, the pipeline was to follow in the track of the Burma offensive, and neither the pipeline nor the Road could reach China unless the campaign succeeded.

All hope rested with American volunteers and equipment. Could the army and AVG volunteers win this campaign? Could we Lend-Leasers solve the pipeline's engineering problems and get oil flowing fast enough to fuel our assault? Nobody was taking bets on our success.

I needed to know all of this, and that thought depressed me. Yu, my faithful Hong Kong host and interpreter, noticed my increasing gloom on the third day when we went to the Chinese consul for my visa. He decided to throw a party for my birthday (six days ago I had turned thirty-four). He gathered a few of his friends for a 14-course dinner at which I managed a passable job with the chopsticks. Between stories about China's warring factions, they taught me the Chinese "finger game." We ended the evening in a Chinese nightclub, where Yu, with an inviting smile, passed me the opium pipe that was going around our table. "When in Rome," I muttered, and gave myself up to the euphoria of the group.

A major blackout was scheduled for the following day, November 14, the day of my departure. Since the ferry stopped running at sundown, I crossed over to Kowloon and met Price and Scott in a hotel there for dinner. The windows and doors were covered with black cloth—only the dining room was lit, a single candle on each table. It was a somber affair.

About ten o'clock we took a cab for the airport. The hour ride was a thrill a minute as we sped through total darkness, hearing the swish of passing cars that we could not see.

At 12:05 A.M. we taxied down the dark runway and took off. For the next half hour, we flew over Japanese-occupied territory with all lights off to avoid becoming targets for their antiaircraft gunners. After a sleepless five-hour flight we landed at a military airport outside Chungking, unlit except for a pinpoint of light marking the end of the runway. I was then driven through the dense fog to the Methodist Mission where, after sleeping until noon, I met P. H. Ho, my official host in China.

Ho was making the arrangements for my survey trip of the Burma Road and my meeting with the essential Chinese and American military officials connected with it. The power plant had been shutdown for several days, so Ho and I dined by candlelight and finished with a game of klick.

The next morning, Ho drove me into Chungking, a mountain town and the capital of the Nationalists. We took the main street, winding along the ridge at the junction of the Yangtze and Kailing rivers, and maneuvering through the press of cars, taxis, rickshaws, bicycles and oxcarts. I was surprised at Chungking's shabby, bomb-pocked appearance and wondered why there were so few military vehicles.

A steady stream of locals lined the road, either carrying their few possessions in bags or pulling them in carts along the cracked and curving roadside. Chungking's 20,000 or so inhabitants were mostly transients, passing through as quickly as they could because of the frequent bombings. Those that stayed functioned as long-term transients.

I saw no paint anywhere. The buildings were made of stucco and plain concrete block, shaped according to the materials at hand and the whim of the builder. Ravaged buildings, bearing evidence of past bombings, were patched with bamboo and mud. Crude straw-thatched huts filled the gaps where more substantial buildings had once stood. Debris, piled near curbs, was to be reused in future repairs. The Nationalist government had commandeered the most usable buildings. Its headquarters and residences stood out because they were clean and well-maintained.

The temperature was about forty degrees, and the day's rare sunshine was a mixed blessing. The usual fog and dampness were always cause for contentment here, since they kept Japanese bombers away. I looked up at a tower we were passing and saw an air raid warden hanging a red-ball, the first warning for a raid. Ho drove along unconcerned.

We reached the financial market. I was carrying a letter of credit and we drew a draft for $50 in U.S. currency. Ho disappeared. When he returned he handed me $700 in Chinese National currency, the current black-market rate of exchange, a rate supported by foreign merchants who needed U.S. currency to meet their import obligations. This was a 35-1 rate of exchange, so I made a 100 percent profit on the deal. This was my first exposure to China's runaway inflation and the resulting black-market cash trading.

We went next to the main airport. Constructed on a gravel bank in the Yangtze riverbed, it is only usable in winter when the water is low. After much loud talk and Ho's explanation that I was an emissary of Dr. Soong, I was scheduled for CNAC's next plane to Kunming.

We then visited the shortwave radio station located underground in a bombproof shelter, and saw another warden adding a second red-ball to his tower. People began pouring out of houses, scurrying to bomb shelters below. Ho and I hurried to one of the caves that honeycomb the hills of Chungking, not waiting for a third red-ball,which would signal an imminent attack. An hour later, a siren announced the end of danger. As we left, we saw that the two red-balls had been replaced by a black flag.

Three days later, after the morning fog dissipated, my plane took off for Kunming, just as red-balls were again being placed on the air raid towers. It was 10:30 A.M.; the Japanese planes would not arrive until noon.

That night, I dined in the military barracks at Kunming as the guest of General Yu Fei-peng. Between the many courses and rounds of gombei, I exchanged information with the Chinese and American military people in charge of the Burma Road. I was frequently distracted by the snarling of the Tibetan wolfhound that guards the barracks door.

Arrangements were made for my supplies and camping equipment, and General Yu offered me his car and a top engineer as my guide and interpreter. On Thanksgiving Day, I celebrated at the American consulate with the other Americans in town, and received a letter from Virginia. I was grateful that everything was all right at home but nervous about my trip down the Road.

That afternoon I visited the electric manufacturing plant a few miles from town. The buildings were so well-blended with the surrounding hills that I was on the grounds before I saw them. Inside, machinery made from parts cannibalized from all over China had been arranged into an assembly line, held together by spit and willpower. Telephones and cables for the army were being made here and, judging from the splintered walls and bomb craters in the yard, it was evident the Japanese knew that. But the people were cheerful and determined, and I was encouraged. If they could do this, we could surely do the damned pipeline.

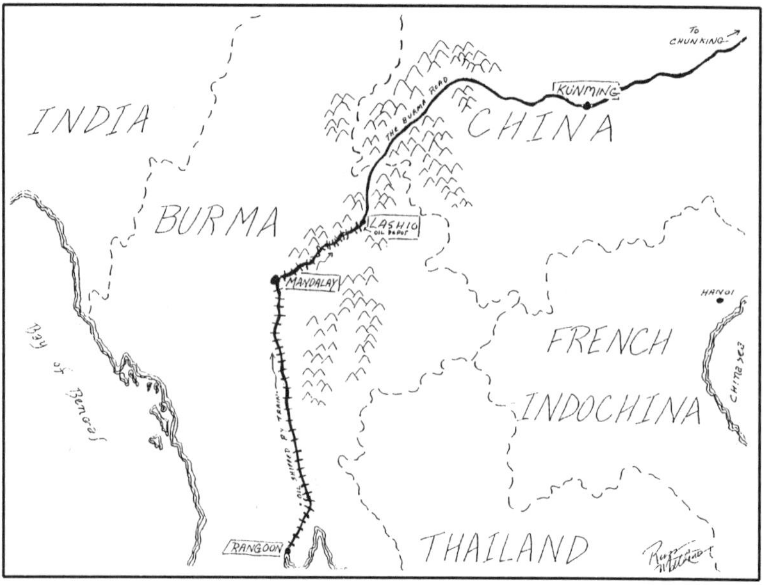
To CHUNKING
KUNMING
THE BURMA ROAD
CHINA
INDIA
BURMA
LASHIO
OIL DEPOT
MANDALAY
HANOI
FRENCH
INDOCHINA
China Sea
Bay of Bengal
OIL SHIPPED BY TRAIN
RANGOON
THAILAND

Chapter VII

On the Road

The next morning, Engineer Hermin Hu and I, driven by Hu's cook-chauffeur, pointed General Yu's Buick out the west gate of Kunming and onto the Burma Road. We moved comfortably along the smooth pavement that winds up and out of the valley through rolling country. We would parallel the new railroad bed until our lunch stop at Lefang.

Excited to be finally on my way, my eyes explored the countryside, as Hu told me about the Road's vital importance to China.

"Now need road for import war materials and export Tung oil. No exports, no worth in China's money. All time true. When peace return, Road used for develop overland trade between China, Burma and India."

After lunch, the pavement runs out and we begin our long ascent into the mountains. We pass a mule-pack train. The spirit of the Road begins to infect me. I wave to the cargo trucks and occasional cars as they pass in a swirl of dust, their gears grinding and straining as they crowd for every inch they can get along this steep and narrow ribbon of road. Because we are loaded down with emergency food, water, bedding and baggage, we are nearly riding on our axles. We make frequent stops to inspect the tires. There will be no replacements until Lashio, still seven days away.

The next several days are a roller-coaster ride on narrow shelves along the nearly vertical mountainsides. There are frequent drops of 2–3,000 feet down steep gorges to the river. The lack of guardrails unnerves me, especially when our driver turns off the ignition and free coasts down the treacherous slopes. The price of gas conservation is high along the Road.

I had been warned about traffic jams at the open-air refueling stops set up every five or ten miles along the Road. Shaking their heads, my American contacts had told me how Chinese truck drivers stop in the middle of the road, hop off while their trucks are being refueled and dash to the lean-to rice shops run by enterprising locals. After passing a number of these without incident, I asked Hu where these interruptions were occurring.

"Ah, no longer that way," he replied. "Road police build parking places next to Road. Now all smooth."

We pass another stop and Hu points. "See him? He owner taxi company, New York. Volunteer. Come here for direct traffic." I see a stocky man with a cigar, waving incoming trucks to the parking area.

The stops have been set up close to caves and tunnels, where drivers can protect their trucks from the frequent Japanese bombers trying to knock out the Burma supply line. Each driver carries a thermos, bedroll and blanket, plus enough cooked rice to last a few days. The bombings are usually heavy and unremitting, and a cooking fire would only draw attack.

Our nights are spent in work camps along the Road; these consist of wooden shacks of low-grade lumber. Each cabin has four doorways but no doors. Wooden gates sit against the wall and these are placed in doorways by the squeamish for keeping out the wind, cold and rain. Crude wall hooks hold our clothes while we eat at long, rough-board tables on attached porches. The beds have wood frames with bedroll mattresses. (In the valley camps, we are sometimes lucky enough to have mosquito nets.) At the barren mountain camps, it is the howling wind that often keeps me awake.

There isn't much to do after the evening meals. The Chinese love to gamble, so I get a lot of practice at the finger game and play a lot of cards. With Hu and the travelling diplomats it's mostly poker. "Red Dog" is the preferred game with the press people.

On the following day, we climb to 8,825 feet and look down on the Hsiakwan plateau valley. The road down the far side seems to vanish into space. In the late afternoon, with the hardest part of the day's trip over, we stop to chat with the drivers of a convoy. Hu explains that most trucks carry an official cargo of gas and war supplies, plus unofficial cargo such as whiskey, cigarettes, cosmetics, or "yellow fish" (unauthorized passengers who pay drivers according to the distance they travel).

"Inflation. Cost of living high," Hu comments. I ask how this gets past custom officials. "Ah, no problem," Hu smiles. "Drivers must eat. Custom officials hungry too. They share."

A little later I see a wrecked truck being stripped. "Also common," Hu explains. "Nothing wasted. Truck abandoned, truck looted."

North of Hsaikwan and the Tali marble quarry we stop to watch the road-surfacing work. I will report to General Yu about the stone mix and asphalt coating being used. Although I see shovels, the coolies aren't using them. They load rocks into baskets by hand, carry them a few steps and then empty the baskets wherever their foreman points. Their tools are primitive and their methods ancient, but the great number of workers makes up for this.

"All workers either Youngs, Olds, Strongs or Weaks," Hu explains. "Strongs and Youngs now in army, so use Very Youngs, Weaks and Olds for road. Also Womens."

Children no more than six, and mothers with babies slung on their backs are breaking stones and doing repair work. Many have goiters. "Common here." Hu is impassive. "Mortality rate high."

To make our passage safe, two coolies climb the rock embankment and dislodge loose stones. The others gaze into space, perhaps seeing the endlessness of their jobs. The Burma Road has indeed been carved by the fingers of China's thousands.

We clear the summit, and for many miles see terraced rice fields along the winding trail. They follow the contour lines precisely, as if laid out by a surveyor. On the approach to Taipingau we are delayed by a traffic snarl. "So sorry," Hu laughs. "Police not get here yet, I guess." I laugh too, as a single village officer shouts wildly at the drivers to leave the rice house and move their trucks.

Near sundown we pass many cemeteries, then enter the village of Paoshan where a telegram awaits Hu. It is from Tan Po Ying, chief engineer of the Road, who will meet us here tonight.

Tan, Hu and I talk far into the night. Exhausted by the time we finish, I walk through the cold night to our camp, and pass several dogs roaming the streets in search of food. I sleep in my clothes and even the flat board of a bed feels good.

In the morning, Tan rides along to show us the Salween gorge and its river, 3,000 feet below. Hu translates as Tan explains the obstacles encountered in constructing the Road here. We wind through rugged mountains and soon reach the east side of the gorge. The river is visible in the distance. I have never seen anything as vast as this great scar on the earth's surface. The Grand Canyon on the Colorado comes to mind. Tan shows us the improvements and alternate routes to cross the river and keep traffic flowing if the gorge is bombed. This promises to be one of the most difficult passages for the pipeline, so I look things over closely.

I feel great admiration for the ingenuity and skill used, especially on the bridges. Cables are well placed and anchored, and are protected from bomb fragments by bamboo sheathing. The road cuts back from the far end, and leads to a pontoon bridge built of wood and empty gas drums. A spare bridge is under construction a mile down the river, along another of the twisting routes that are typical along the gorge. All routes are carved out of the face of nearly vertical mountainside, and trucks in the distance look like ants crawling down the mountain.

Even with our modern equipment, this section of pipeline isn't going to be easy. Since no photographs of the Road are allowed, I take the time to draw sketches and take copious notes on the area.

The trip over the summit brings us to rolling wooded country and into Lameng for a late and hasty lunch. Aiming for Manshih by nightfall, we drive rapidly through the last semi-barren set of ranges. Far to the south we see an endless expanse of mountain ranges which gradually diminish toward the Burma frontier.

We soon drop down into a warm and humid valley of rice fields, and stop to take pictures of a boy riding his water buffalo. The boy, speaking a dialect different from Hu's, cannot answer our questions but seems very amused by us. Hu tells me that the wild life in this area was long ago killed off for food. The local diet now consists of rice and a few vegetables. We pass through dense growths of bamboo, and see huts, fences and foot bridges made exclusively of bamboo.

At Manshih we are invited by "the Sawbwa" to spend the night as his guests. His is the only modern building in town so we jump at his invitation. Following dinner and a round of the finger game, his Burmese wife interprets as the Sawbwa tells of his tobacco and sugarcane crops and of the 200,000 tung oil trees that have been planted for export. There is also a small silk industry here and undeveloped deposits of coal, iron, silver, lead and zinc. I wonder if the Sawbwa is a Nationalist sympathizer and if any of his income goes to defending his wealth.

The following day, as we drive the last section of the Road to the Burma border, we notice warmer weather and a gradual drop in altitude. I make frequent inspection stops. Poor road surfaces and drainage problems has caused a lot of traffic trouble here during the last rainy season. We see many groups of workers breaking stone and smoothing the grade so that asphalt can be applied.

The pocked surface keeps our speed down to twenty kilometers per hour, giving me time to notice that many of the workers are women. Their children play at the side of the road, where piles of rock act as temporary guardrails on the sharp curves. I worry about the children, but am comforted by any kind of guardrail after the many times we have come within inches of an accident or a plunge over a cliff.

It is nearly noon when we stop at Wanting for an inspection of the truck parks and fuel dumps. Since our Buick doesn't have Burmese plates, we change cars for the trip to Lashio. Only one car is available. It is a tight squeeze for all of us and our gear. The car, a Chevy, limps into Lashio after dark, the victim of one flat tire and three broken springs. Hu's driver has done amazing things with pliers and a roll of wire.

Hu and I fly to Rangoon, then travel 200 miles by car to Bhamo on the Irrawaddy River, glad to escape Lashio's malaria. This 200 miles is rough going, resulting in several more broken springs. We creep the last sixty miles in the dark, through tiger-infested country, and at night sleep fitfully in the car.

At the tankage installation on the river, I inspect and inventory the available facilities. This completed, we start the return trip to Lashio through the mountainous jungle. Travel is limited to one-way traffic for the sixty miles east of Bahmo, alternating direction every twelve hours. We just make the noon deadline for east-bound traffic and drive through.

The road to Lashio is paved, so we settle back for the rest of the trip. At our Lashio hotel, Hu and I draft a report for General Yu of this week's travel. I make copies of this report for later inclusion in my reports to the China Defense Supply and to Chiang Kai-shek.

The next afternoon, on the plane to Rangoon, I run into Bousfield. His urbane humor lifts my spirits, and we joke about conditions on the road. The flight, seeming like good ol' days with Hal Sweet aboard as our pilot, relaxes me. Only now do I realize how tight and aching my muscles have gotten on this trip.

Bousfield and I set up housekeeping at Rangoon's Strand Hotel and go out for a relaxing game of golf. It is hard for a midwesterner such as myself, accustomed to cold winters, to believe I am playing golf on November 30.

Along the road, the primitive living conditions had kept me so busy dealing with immediate survival needs that I had thought about little beyond my observations and notes. In Rangoon, the strict blackout brings back the tensions of impending war. But I manage not to think about war as I trade news with the many Americans and

British in town. Several are oil men, and I make arrangements to see them in Chungking on my way back.

The next week is a whirl of visa applications and meetings with American, British and Chinese oil people. We review the entire supply position and make our plans. The meetings go slowly because of the wide differences of opinion between the Eastern and Western groups and the seriousness of the supply problems. To get the military point of view, Colonel J.J. Twitty is included in the last meetings. I learn of many problems that have not been reported to superiors back home, and therefore not passed along to me.

It takes me a full day to go through the docks and godowns (warehouses) along the Irrawaddy. All Lend-Lease goods for China are unloaded here, and distribution problems have caused a bottleneck. Containers are piled everywhere, waiting for available trucks or rail space. Little use has been made of river transport, and rail movements are slow and uncertain.

This inefficiency infuriates me. Scarce and precious oil sits here unused, while along the Road trucks are making do with vegetable oil as fuel. I have avoided pulling rank so far, but this is too much.

I arrange with the Irrawaddy Flotilla Company to move large quantities of gas down the river to Bahmo, and check out the new GM assembly plant to see how the new trucks for the Burma Road are coming along. With the congestion ending and the promise of more trucks soon, I feel a little better. There is little else for me to do here, so I check on my passage back.

Calls at various Chinese agencies reveal a gloomy picture. When I am finally given my Hong Kong visa, the Burmese officials express doubt that I will get to use it. "Events of the next few days," one remarks darkly, "will determine that."

I hurry through visits to plants, the refinery, through meetings with Burma Oil officials—trying to get promises of help with labor in north Burma. My urgency increases as I am shown how the refinery has been rigged to blow up in case of attack.

Thank God I have reservations for Monday's flight to Chungking! What with the blackouts, the siren just outside my hotel window, and my need to get some fast action going in Chungking, I am chafing to get out of here.

Saturday is "hurry up and wait." I put my notes together and send a final cable to Washington. Then it's play time: the Rangoon Turf Club for the Governor's Cup race. Sunday is more of the same, playing golf and sight-seeing at the Shwe Dragon Pagoda with its famous gold-leaf dome.

As I finish packing this evening, a great longing for home overtakes me. I wonder what Virginia is doing at this moment. Here in Rangoon it is almost midnite, Sunday, December 7. I stare at my calendar, willing my overloaded mind to compute the time differences. Let's see...first deduct a day at the International Dateline near Manila. Then add five hours for Hawaii time, making it 5:00 A.M. today in Honolulu. Add another seven hours to get to Eastern Standard Time, making it Sunday noon at home. Sunday night in Rangoon but only noon in Indiana . . . .

The kids are probably out playing by now. Virginia, ever gregarious, is probably either playing bridge with friends or practicing on the piano. I think about her warm blue eyes and prematurely grey hair as I continue staring at the calendar. But no matter how hard I look, today is December 7 in Rangoon and I cannot leave until the morning of December 8.

Chapter VIII

A Change of Plans

I arrived early to the airport, worn-out from a sleepless night. The damned speakers were blaring through the ticketing office and everybody seemed frozen in place. "Well, shit," I said to myself, "another bureaucratic snafu. I'll probably be stuck here all day." It was a few moments before I stopped feeling sorry for myself and focused on the words:

> *Word has just arrived from headquarters of the Flying Tigers in Mandalay, that the Japanese have attacked Pearl Harbor. Repeat: Japan is now bombing the United States' naval base at Pearl Harbor. Extent of the damage is not yet known. The United States is now at war.*

My breakfast rolled into a ball. Out there on the immense blue ocean, time had run out on me. But Pearl Harbor? Everyone had been expecting an attack in Southeast Asia. Instead, the Japanese had flung themselves on the American battle fleet thousands of miles across the Pacific.

This daring attack had begun only minutes before and few details were known. Panic rumors began flying: the Japanese fleet ten minutes off the California coast; San Francisco expecting an air raid; a periscope sighted off the northern California coast. I ignored all reports except those coming through the Flying Tigers.

In the agonizing hours that followed, we learned that the Japanese had also attacked Guam, Wake Island, Hong Kong, and the Philippines. Our last report from the Tigers announced an attack on Malay.

Our plane took off on schedule, in spite of the confusion and fear surrounding us. During the flight we received radio reports confirming the bombings of Manila and Hong Kong. The reports were sketchy, and most of the flight-talk was speculation on what the hell would happen next.

At the Chungking airport that evening I learned I could forget about that Clipper ride from Hong Kong to the States. My plane had been strafed and burned at the dock, and CNAC had lost five planes on the ground.

With the Japanese busy elsewhere, Chungking was as safe a place as any. That night I returned to the Methodist Mission where I ran into Emil Scott and several other CNAC pilots. They were resting from their Hong Kong-Kweiking shuttle, now the only escape route, on which they were evacuating key Hong Kong people. I had another sleepless night, trading news with arriving and departing pilots as they kept up steady trips to Hong Kong with the few planes left to them.

Midway was hit the next day, and the following day, we received incredible news: Britain's newest battleship, the *Prince of Wales*, and the battle cruiser *Repulse*, sent to the Far East in November to deter attack on Singapore, had been sunk by aerial attack off the coast of Malay. The British had lost command of the sea in Southeast Asia.

On December 11, Germany and Italy declared war on the United States.

Even the Fourth Army Headquarters was panicking now, reporting that the main Japanese battle fleet was 164 miles off the San Francisco coast. Chungking was in pandemonium, the situation was shifting with each new report.

"The Chinese are overjoyed," Scott told me on one of his layovers. "The Pearl Harbor attack feels like salvation to them. Chinese officials openly congratulate each other, acting like some great victory's been won! Yeah, China feels like it has allies at last now that we've entered the fight against Japan . . . the good old U.S. can be counted on to finish it off."

He was right. I'd watched Chinese crowds rejoicing in the streets for the first few days. One American watching commented, "Pearl Harbor day in America is Armistice Day out here."

But as reports of the crumbling Western position reached Chungking, discouragement replaced the Chinese euphoria. Chiang Kai-shek was busy angling for power in the Western alliance to confirm China's importance in global strategy-making.

I haunted the British and American military headquarters, hoping to make radio contact with London or New York for instructions. Nobody would even respond to me.

British and American military headquarters were caught in a struggle over areas of command. Clerks were rushing in and out, followed by shouts of "Get me ten copies of this at once!" or "Get that secret file out of the safe!" and "Where the hell is the Orange (Blue, Green, Yellow) Plan?" Everybody was furiously smoking cigarettes and passing me on to someone else.

Between ringing buzzers and telephones, crowded rooms with clerks banging away at typewriters and people jumping up just as I started talking to them, I finally managed to see General MacGruder. I brought him up to date on my trips in China and Burma as quickly as possible and delivered Colonel Twitty's report on the pipeline. He was cool and would make no commitment until the pattern of the military situation became clear.

Damn. Added to everything else is now a tough selling job to MacGruder. He has some financial say on the project and Lend-Lease might want his approval before proceeding. Might. I had no way to know.

I figured with the Japanese moving in, the pipeline had become even more important now, and plans to complete it would be hurried along. I'd just have to fly blind for the time being, sticking as close as I could to the original assignment until I got other instructions, maybe through Chungking Shell.

Before I could leave Chungking, I had three goals to accomplish. The first, of course, was to arrange passage out, but that was meaningless until I obtained pipeline approval from General Yu and his cousin, Generalissimo Chiang Kai-shek. Then there were the many meetings with American and British diplomatic and military personnel, and with British, Chinese and American oil people. I hoped like hell they weren't all going to be like the MacGruder meeting.

The plan, up until the Pearl Harbor attack, had been for the Americans to take over the crew working on the Road and pipeline in China and to eventually meet up with the British at the Burma border. The Brits were then to take over the pipeline all the way to Kunming, where the Chinese Nationalists had put their military headquarters because of its distance from Japanese bomber bases. I'd just have to assume this split until otherwise notified. No matter who did what part, I'd still have to coordinate with the military and the diplomats.

The meetings with the diplomats alone would take weeks. It was their job to contact the province heads about the requirements along the Road. With each province boss, cooperation and assistance had to be worked out regarding labor and truck stops in their areas. The truck stops would be the property of the local bosses, but controlled by the Chinese army as long as needed. Some delicate negotiations would be required in the provinces with communist sympathies.

The only good news was that the Japanese had let up on their bombings of Chungking.

Chapter IX

The Dog Fight

I moved to Shell House, opposite the airport, where accurate reports were now coming in over Shell's private cable system from London, New York, San Francisco, Hawaii, Australia, Venezuela, and occasionally from the French underground. Governments around the world ensured that information was constantly fed to the oil companies about worldwide conditions, because if oil was going to reach necessary destinations, we needed to know which shipping lanes were being torpedoed. What news didn't arrive over Shell's coded cable system filtered in from the daily military cables to army headquarters in Chungking.

Chungking Shell's people were relieved that the U.S. was now openly in the war. To the oil industry, it meant better control and utilization of supplies because they could put more people into transportation. Despite the terrific tanker losses to torpedoes, it made their job easier. Temporarily doing away with Congressional interference through antitrust suits, the secretary of the interior was assigned responsibility for the oil companies. Consequently, they were able to pool resources and considerably reduce the amount of shipping requirements.

I smiled, thinking about how Bousfield and the rest of the oil hierarchy must be pleased with this governmental concession. They were probably already planning ways to keep this in effect after the war ended.

But the government was not doing any favors for me. A policy decision was made to stop shipping supplies for the pipeline: Europe first. The stuff earmarked for China was being diverted!

Loaded with pipeline supplies for China, a ship in Cairo was intercepted and sent to Russia to improve movement of Russian oil to Allied forces. Meanwhile, Chinese oil supplies lost through enemy attack were not being replaced. Orders were that the pipeline was to go ahead, jerry-rigging where necessary.

At first, I was grateful for the supplies that had flooded into Rangoon about the time of the Pearl Harbor attack. But soon, I realized they represented another headache. Britain was trying to grab them away from China.

Hostilities between China and Britain surfaced during Chiang's first meeting with his Western allies. Imperial Britain didn't share the American interest in a strong China after the war, and China feared and disliked the British, who had been the first to penetrate Chinese territory and who had upheld unequal treaties during the colonial era.

In the desperate hours of December 1941, Britain's immediate concern was to hold Burma as the last barrier to India. Britain had little respect for China's government or military capacities. They also felt that the Lend-Lease stockpile in Rangoon would be wasted on the Chinese.

China was equally interested in the defense of Burma. They advocated keeping the thousands of tons of materials already stockpiled in Burma, as well as keeping open the flow of Lend-Lease supplies coming through Rangoon.

At the Chungking conference Chiang's primary concern was this struggle for Lend-Lease goods and control of the newly arrived American Volunteer Group based in Burma. China had title to both, but the British argued that they could make better use of them.

The American view was that Britain's position in the Far East was feeble, mainly a show of power using scarlet uniforms, bearded Sikhs for police, parades, and sharply-executed parade drills. Britain's real reason for opposing Chinese troops in Burma was concern over China's traditional claims to Burma.

The British wanted to sidestep China. They sought to maintain the status quo for the British Empire. For the Americans, China was focused on as a future world power.

The word around Chungking was that the American chiefs of staff considered the linkage of Burma and China and the opening of the Burma Road as indispensable to victory in the war. Secretary of State, Henry Stimson said that the British had reacted in a tactless and peremptory way toward China. Rumor had it, he was worried that the Europe-first strategy would result in China's collapse from discouragement.

Roosevelt and Churchill argued over the contribution China could make to the war. Roosevelt insisted that England stop treating Chiang as a barbarian chieftain. Roosevelt believed the days of colonial empires were over and that a fundamental East-West cleavage would result if the Western world didn't stop treating Asians as inferiors.

Things seemed to be lining up with the United States on one side, Great Britain on the other, and China, in the middle, ferociously guarding its bone. That bone happened to be the supplies I needed to get my job done.

"How's it going?" Scott asked one night, while we were getting drunk together.

"Who the hell knows?" I answered sourly. "I'm just a mongrel pup, watching the three big dogs fight over my bone."

"Huh?" Scott answered, finishing off his drink. He'd just gotten off forty-eight-hour duty and was busy pickling his brain so he could sleep. My bone-metaphor was too much for him. I explained:

"Chiang barks, 'It's mine!' And technically he's right."

"'A waste!' Britain growls, moving in for the grab."

"'Hands off!' the Americans growl back, hackles up, and the three of them go round and round again. So I don't know how it's going. Ask me again when the dog fight's over."

I poured us each another drink.

Scott emptied his glass in a gulp. "Oh, yeah . . . *those* dogs!" he snapped, his red-rimmed eyes lighting up with understanding. " Hey, don't worry about it, Hall. Our side'll never let those limeys have your stuff."

I drank to the hope that he would be right.

In spite of my toasts, a few days later the stockpile of American goods at Rangoon was diverted to the British for the defense of Rangoon, now under imminent threat of bombings.

The high-level jockeying continued, making my job a nightmare. I was going crazy, keeping one set of notes according to plans laid before Pearl Harbor, and another set that included the changes made since then. While I shuffled papers, the war went on.

On December 23, Guam and Wake fell, after 100 years as a British stronghold. Hong Kong surrendered on Christmas Day. The Japanese had landed in the Philippines and were driving the American-Filipino forces into the bottleneck of the Bataan peninsula.

Shared with the Americans and British in Chungking, Christmas dinner was solemn. We drank a bit and talked little, everyone absorbed in their own thoughts.

As I picked listlessly at the roast duck, I thought about the turkey and Christmas tree I wanted to be sharing with my family. Virginia had probably taken the kids to her parents' home in Carlinville, Illinois, for the holidays. I wondered what presents the kids were squabbling over this year. I missed even their bickering. The more I drank, the farther back my thoughts took me. I remembered the frat party at Washington University where I'd become Virginia's blind date, to help a fraternity brother who had ended up with two dates. God, how excited I'd been about her! Even after she had agreed to see me again, my nerves continued to dance. The strawberry taste of her lipstick and her jasmine perfume came back to me as if it were that very evening.

It was a few years later before we were married. It was only then

I'd found out what a trooper she was. We shared a house with two bachelor co-workers on my first job for Westinghouse. Virginia cheerfully picked up after all three of us and fed us our favorite foods.

As I continued to pick at my duck, my childhood memories of my home in Webster Groves suddenly jumped into my mind. I thought of the wonderful Christmas goodies baked by Annie, our live-in maid.

Like Virginia, Annie was always laughing. A big, black woman, she had been a second mother to my two brothers and me. Annie had loved us with a pure and simple devotion I had never known since. I was missing her tremendously.

Our house had always been full of neighborhood kids. I had never been sure whether it was because of Mother's graciousness or Annie's cooking. But I knew it was Mother who had taken our side against Dad's strictness. I was missing her more than ever.

I was missing our cats Mike and Ike. They would follow me to the movies. In the midst of reminiscing, I passed out. I have no idea how I got from the Christmas party to my hotel room, but my head hurt for days.

♦ ♦ ♦

I was gradually working my way up the hierarchy of Chinese officials and, on New Years Day, finally submitted the pipeline plan to General Yu. It took another ten days to gain an audience with Chiang Kai-shek himself.

In spite of his pressures and disappointments, Chiang and Madam Chiang received me graciously, without the usual Chinese formalities. We sat in a circle, including one junior Chinese general to answer military questions. Soon we dismissed the official interpreters because they were not translating the technical information in correct detail.

I asked, "Would the generalissimo like me to explain what we can do for him?"

"Yes," Madam answered, and took over as interpreter.

I put my maps on the floor in front of Chiang's easy chair and we began, Madam translating his Chinese for me and my English for him. This continued for several hours, until we were sure all of us thoroughly understood each other's points.

Madam Chiang sat pliantly straight through this, remaining graciously patient through tiring repetitions and difficult translations. Chiang's concentration was impressive. His black eyes shone with interest, yet never lost their calm as Madam translated technical details in answer to his many questions. Hard as I tried, I could not find simple analogies for the hard facts of this project. Time after time, Chiang asked about the availability of supplies. Time after time, I reassured him.

Slow and tedious as this process was, the Chiangs remained pleasant. After our business was concluded, we drank tea and chatted about the United States. The generalissimo's parting remark was, "I hope your superiors will accept what you have outlined for China and will arrange for the supplies shortly."

After more than a month, I had completed my goals in Chungking and couldn't wait to get the hell out of there. But even with Soong's intervention, passage out had not yet been arranged. I filled my time by lecturing China's Transportation Control Board on "The History of Pipelines and Their Possible Application in China," while renewing my inoculations, and pestering Ho about arranging a flight out. This trip had become the longest two months of my life.

At 4:30 A.M. on January 19, I said good-by to China and was on my way to Calcutta.

♦ ♦ ♦

This was my first visit to India, but I was too intent on getting home to pay much attention to any new surroundings. In a blur, I shuttled from plane to plane, enduring inoculations and excessive customs formalities in Calcutta, Karachi, Basra, and Cairo. Finally, after two stops in the Sudan and one in South Africa, we took off for Bermuda and the States. On February 15, I got off the plane in Baltimore and took the train to Indianapolis. By 9:00 A.M. the next morning, I was covering Virginia with kisses and laughing at the squeals of John and Kathryn as I crushed them in big bear hugs.

Chapter X

Homecoming

Virginia drove us home from the Zionsville station. My eyes devoured the familiar route as I fielded John and Katie's questions.

"Were you bombed, Dad?" asked wide-eyed Katie.

"Did you kill any Japs?" inquired patriotic John, who was disappointed to learn I had seen no combat. I think I went down in his estimation, for handling only valves and paperwork instead of weapons. But Katie still gazed at me in awe of her dad—the war hero.

As we neared the house I cautioned Virginia, "Pull into the garage slowly. Don't make any noise." The Zionsville plant superintendent who lived behind us had a nosey wife. I didn't want the neighborhood alerted just yet that I was home.

John helped me carry in my luggage while Virginia and Katie got breakfast. And what a breakfast! Big slabs of ham and biscuits along with a mushroom omelet, juice and coffee cake. The familiar flavors did a lot to dissolve the weariness that had settled into my bones. I longed to top it off with coffee and a cigarette, but the kids' excitement wouldn't permit it. Today was our private Christmas party.

"What'd you bring us?" they shouted in chorus, mouths still stuffed with eggs.

"Oh, a thing or two," I grinned. "John, hand me that flight bag, will you?"

He vaulted over the back of the couch and in a moment stood before me, flight bag in hand. I felt like Santa, pulling out trinkets one by one with great ceremony and delivering a little story with each gift. John lit up over the ivory tusks from India until he saw the water pistol I had picked up in the Baltimore train station.

"Bang! Bang! Take that, you dirty Jap," he shouted, running around the house and dropping to the floor in a writhing heap as his imaginary enemy plugged him with a lucky shot. Virginia grimaced at me, then said to him sweetly, "You may fill that with water if you'll take it outside." He was gone in an instant and we sighed at the sudden calm.

Katie was absorbed with the assortment of boxes from China which she had surrounded with her tea set and dolls. She was lost in her own world of make-believe, doing a creditable imitation of a Chinese tea ceremony. It suddenly dawned on me that the kids must have missed me deeply, and that Virginia had probably soothed them with stories about China and the other lands I was in.

Virginia loved the Chinese antiques I had brought her—belated Christmas and Valentine's Day gifts, purchased with love and good connections. She especially liked the ivory statuary I had managed to buy through interpreter Ho in Chungking. Ho had arranged to have these treasures packaged and labeled in Chinese, with Soong's name prominently displayed to guarantee a clear path through customs.

I finally got my coffee and was settling down to tell Virginia about my trip, when the doorbell rang. Virginia put down the army gloves she was knitting and went to the door.

"Oh, Mrs. Hall. Is everything all right?" asked the plant superintendents wife. "I heard shouting and wondered if.... Well, Mr. Hall, when did you get back?"

That was the end of our privacy. Soon everyone in town was calling or stopping by to invite us over. Virginia made excuses, putting them off for a few days because of my exhaustion. I telephoned her parents and mine, promising visits as soon as I filed my reports in Washington and New York.

That night, soaking up Virginia's warmth and the sounds and smells of home, I slept deeply and peacefully. But sunrise found me wide awake, refreshed and grinning at the sounds of birds in our backyard. This tranquility lasted until 8:00 A.M. when, over my first cup of coffee, Virginia called, "Telephone, Jack. It's New York."

It was Alexander Fraser, head of U.S. Shell. Our nosey neighbor had worked fast. Fraser inquired about my health before announcing I was to report to New York in a couple of days. He was as anxious as I to quickly get my top secret reports into the right hands. He expressed his relief over the success of my trip and mentioned that, with things heating up in Europe, Shell was eager to have me back at work on the English pipelines.

In spite of the extensive work I had done on that project last year, the urgency of the Burma Road pipeline had wiped the English project from my mind. I ordered my weary brain to switch tracks as Fraser and I discussed the situation for a few minutes.

"Take a few days for yourself first, though," Fraser offered. "Visit with your family and friends. Rest up from your strenuous trip. Have that tooth taken care of. We want you tough and lean when you get here."

Virginia took it well. She understood the responsibilities of a company wife, and seemed to be getting used to my absences. Maybe too used to them, I thought, remembering my excited call to her from Baltimore. I had babbled into the phone, "Virginia! I'm home!" After a long pause she had answered, "Who's this?" She still hadn't lived that one down.

The next two days went by fast, taken up by dental appointments and a hello trip through the Zionsville pipeline plant. There was no time for the many friends who wanted to throw welcome-home parties. I was one of the very first to return from a war area. I knew they were eager to hear about my trip, so I agreed to give a talk before the PTA at the township school.

It was an amazing turnout. Everyone in town showed up, including state senators and representatives, local councilmen and politicians. There was even press coverage from Indianapolis. I met the whole state that evening, making some important connections in the state capitol and ironing out right-of-way problems for the Zionsville pipeline. The mood was patriotic. Anything the government could do to help supply oil for the war effort would be done.

The next morning I caught an early train into New York, stored my luggage in a locker at Penn Station, and hopped another train for Washington. By noon, I was in the China Defense Supply offices, unlocking my diplomatic pouch and delivering my report. CDS got my second set of notes, reflecting the changes since Pearl Harbor, and notification that I could be reached at New York Shell if I was needed.

Soong's first assistant seemed surprised when I handed him $1,000 cash—expense money I'd drawn in China and not needed—but willingly signed the receipt I insisted he provide. Knowing what I had learned about the business practices in China, I knew that receipt might be my only record that this money had been returned to CDS.

I was sure they had assumed I would use my expense account for personal spending, but I'd kept it clean. That hadn't been hard. I'd taken along money of my own and built it up considerably with my Red Dog winnings. It had been made even easier by my belief that "when-in-Rome" didn't apply in business. I still worked for Shell so I played by Shell rules. I didn't want to owe the Chinese anything and this settled my account with them.

The next morning, I delivered the receipt to the Shell Treasurer in New York and headed for Fraser's office on the 41st floor. Vice President McGraw and Operating Manager Syd Smith of the Burma Road pipeline were there waiting for me. Once again I unlocked my pouch, this time delivering my first report based on my original assignment.

While Fraser, McGraw and Syd looked over my notes, I unchained the pouch from my wrist. A giddy sense of freedom flooded me. I had shed all the papers from my China mission. That meant I no longer needed to maintain the constant vigilance of guarding secret papers.

Fraser broke the silence. "So Chiang says he'll take our pipeline, does he? White of him, considering it will probably save his neck. But don't think I'm underestimating what you went through to pull it off, Hall. Damned good job. We won't forget it."

Syd squirmed, no doubt remembering that only his fear of flying had kept him from getting this assignment. He was a brilliant researcher. Mastermind of the Burma project, he had been the driving force of its development. He also had been the logical candidate for the trip. The only other possible candidate was Hugh Dischinger, Superintendent of the Chicago pipeline, but he was not as experienced as Syd or I. Syd knew that his inability to fly had cost him this assignment and limited his opportunities for promotion.

"Looks like you've covered all the bases," Syd generously acknowledged, flipping through my report.

"All I could," I answered, "but we'll have to straighten out the can of worms created by Pearl Harbor and the reassignment of supplies."

"Damn!" Snapped Syd, punching the arm of his chair. "After all those months of putting together specs and finagling with suppliers for deliveries—the accounting on CDS funds has turned into an unholy mess. It'll take a couple more weeks just to figure out what's left in the project budget. Then we'll have to trace what's gotten there and what has to be reordered. Better plan to be here for a while, Jack. There's going to be a lot of work left after we translate your report into supply orders.

"From now on, Hall will just be a consultant for the Burma line. We're moving him back to work on the English pipeline problem."

Syd's face fell slightly before he nodded. Fraser had spoken and that was that. I wasn't too happy myself. We all knew what 'consultant' meant. There was no way Syd could pull the Burma pipeline out of the fire without a lot of help from me. I now had two full-time emergency projects to juggle. But I couldn't complain. Active duty was the alternative. A foxhole wouldn't be any easier, and these pipelines were more important than killing a few enemy soldiers.

Syd got to his feet. "Okay to pick Jack's brain for the rest of the day, Mr. Fraser?"

"Sure. But don't make any plans for tomorrow, Hall."

Syd and I left Fraser's penthouse office and took the elevator to Syd's office on the 36th floor. The RCA Building—clean, immense—still startled me after the squat filth of Chungking.

"Okay, Jack. I'll set you up as soon as I locate an available office. The rest of the team is scattered between four or five floors . . . most on double assignment, like you, so it's hard to keep track of them. I'll have a road map made up so you can find them. Meanwhile, let's go over your report and clean it up for typing. I'll need your help in getting this project back on line."

It was evening before I got back to my hotel. There was a message from Fraser waiting for me at the desk. I tried his office and his home, then connected with him at his club.

"Ah, Hall. Just get in? Long day for someone just back from the trenches. You need a break. I've arranged a breakfast date for you tomorrow with someone I think you'll find very interesting. Tell you what, have a shower and some dinner. By the time you've finished, I'll have a messenger there with the particulars. Enjoy."

Late that night, with Fraser's note in my pocket, I caught a military plane to Washington. In the morning I went to the Georgetown address I'd been given, which turned out to be an elegant mansion on a quiet street. I was surprised to see last year's car models in this expensive neighborhood, until I remembered that new car sales had been banned on January 1st.

The man who let me in was no ordinary butler. He searchingly looked at me, with too much apparent awareness. I thought about Fraser's vague mention of a ruckus between the FBI and Donovan's secret service group. I realized what a high level I was working on as the butler issued me into a luxurious morning room and presented me to General Bill Donovan.

Donovan was intense and charismatic. He was heavily built, but seemed powerfully athletic. He was dressed in a quietly expensive business suit. His hair was not quite silver and his skin gave off a deep ruddy glow. He looked at me with his clear deep blue eyes and seemed to size me up in the moment of our handshake.

"Hear from the Brits you had a rough time with one of our people in Bermuda," he said with a genial smile.

Bermuda? His people? Then I remembered the immigration officer I'd battled with on my way home. I was fed up by the time I'd reached Bermuda, especially with the diplomatic pouch that was chained to my wrist day and night when I carried important papers. Because of my top secret clearance, I'd been tapped for courier duty throughout the trip. Sometimes I carried papers only short distances, such as from Chungking to Kunming, but often I'd been manacled to that pouch for ocean-going flights. I'd left San Francisco carrying papers for Hong Kong. On my roundabout return flight I'd brought back a number of documents for Soong from Chiang's generals, as well as papers for the Netherlands ambassador to the U.S.

I returned Donovan's smile as awareness dawned.

"Oh," I said sheepishly. "All I knew was that one more immigration officer was putting me through bureaucratic grief. By then, I'd gone through what seemed like thousands of immigration procedures, so when this guy in Bermuda insisted on inspecting my sealed pouch, I flipped. Even if I'd known I was dealing with a secret service man, I'd have acted just the same."

"That pouch contained papers about our Burma Road plans and the transfer of control to the British. I was damned if I was going to forego diplomatic immunity for some immigration yobo who'd decided to flout the protocols of international law!"

Donovan laughed. "You did the right thing, much to our agent's surprise. He's used to docile cooperation. He was a little startled—pleasantly, I might add—when you staged a scene and demanded to speak to his supervisor."

I had forgotten that part and felt a little embarrassed at the reminder, but Donovan quickly put me at ease.

"The way you handled that, plus your government connections, are why I asked you here," he said.

His "butler," who seemed to be the only other person in the large house, served us breakfast while Donovan pumped me about the various countries I had travelled through. Had anyone seemed too curious about my assignment? Had I noticed anything unusual? Had any civilian Orientals been exceptionally cordial to me? I answered his questions, then told him about the problems of getting materials and equipment into China, explaining the great need to speed up construction. The Far East needed all the help we could give to keep the Japanese at bay.

"Yes," he said. "I'm having a little trouble there myself. But I understand you're being taken off the Burma pipeline and assigned to the British problem."

He seemed to know everything.

"True, but I'm determined to finish what I started there. I gave the generalissimo my pledge that the Burma line would be finished. I can't drop it just because I'm stuck behind a mountain of English paperwork."

Donovan looked at me thoughtfully, then said, "I understand you had ROTC training in high school and college. Gave up your Reserves status to work for Lend-Lease, right?"

"Yes." I wondered where he was going with that.

"I also understand you were nominated for West Point by your ROTC commander at Washington University. Mind telling me why you turned it down?"

"Simple. It was the depths of the Depression. My four years at college had been a hardship for my parents. And my brothers, Dick and Lee, were in college by then. I needed a job, not three more years of training. Besides, I was engaged."

The trace of a smile crossed Donovan's face. "Well, you obviously mastered the military mindset, and I'm sure that mastery was indispensable to coordinate things in China. You speak their language and you understand how they operate. That means you know how to translate business and government requirements into terms they can appreciate."

"Those same skills will be needed to put the English pipelines into shape to supply troops, ships and planes for the European theater. Tell you what: if you'll agree to be field man for that job, OSS will back you up on both the English and Burma pipelines . . . run interference, so to speak, when you hit snags with higher-ups. I could use a man like you for liaison work. Interested?"

I agreed and for the next couple of months in New York I shuttled back and forth between my Burma office on the 37th floor and my England office on the 38th. I spent a lot of time following various files around to offices that shifted to wherever project personnel camped out. It was hard to stay organized in this scatter and hard to connect with the people I needed.

With so many overlapping assignments, I never knew when a key man would be out to lunch, in a meeting, on a trip, or simply on a different floor of the building working on another assignment. Long weekends at home, with ordinary people, helped keep me sane.

War mobilization went into effect in February. Our weekend socializing included lonely wives and older men. Young husbands went off by the millions to combat and women took jobs in war industries and military support units. Virginia rolled bandages and knitted army gloves. Rationing was introduced. We went easy on coffee, butter, sugar and meat, and we made few weekend drives because of fuel conservation.

Zionsville, less than thirty miles from Indianapolis, was a different world. The pipeline works was the only industry, surrounded by rich farms and country estates. The Pike Township School had twelve rooms and twelve grades. Scholastically it was excellent, remarkable for a country village of only 600 to 700 people, but we had had a recent influx of people from Indianapolis.

Zionsville was becoming a suburb for executives and state officials. These new immigrants from the city were our neighbors and friends. Their kids were friends with our kids, and the distinction between city and country folk was becoming blurred. Many of the area "farmers" were important people. One of our neighbors had invented a machine to row-plant (rather than scatter) soy beans.

We lived on company grounds, so Katie and John had no next-door friends to play with. This didn't seem to trouble them. Katie seemed content with her after-school athletic activities, Girl Scouts and 4-H. John was fascinated with the pipeline operations and anything mechanical. He sat for hours at our living room window, closely observing the huge oil trucks that passed by twenty-four hours a day.

John always asked to come along when I went to my plant office. The pumping station, railroad loading dock and big equipment such as ditch diggers, bulldozers, fork lifts and the like, fascinated him. But his new obsession became dangerous. After school, he and Katie would visit their new playmates, the plant workmen, and their biggest adventure became sneaking onto the plant grounds. Even Virginia's many applications of the table tennis paddle did not deter them, nor were they afraid to try outwitting the Doberman and German Shepherd guard dogs.

But the war helped. Soon the kids were engrossed in growing a Victory Garden and raising hogs, chickens and a goat for 4-H. They helped Virginia with the canning, collected scrap metal for the war drive, and picked milkweed pods for stuffing in life rafts and jackets. There were few able-bodied men left at home, so John began doing occasional odd jobs on local farms. With the farmers' permission, I taught him how to run their farm equipment. While Katie practiced piano, John and I built crystal sets. There were barbecues, church fish fries and, as the weather warmed, swimming outings and golf lessons for the kids.

It was a happy and busy time at home, peaceful except when the kids' pigs got loose or their goat got into the house. Virginia almost killed that goat after it ate the tablecloth her grandmother had made, but the kids' tears softened her as they promised to check the lock on the gate "*three* times *every* day, Mom."

The kids went around the house singing "Praise the Lord and Pass the Ammunition." In May, the Women's Army Auxiliary Corps was established and Corregidor fell. "Blues in the Night" and "The White Cliffs of Dover" played constantly on the radio. In June, I kissed the family good-by and set out for England.

Chapter XI

A Wing and a Prayer

I thought my years in ROTC and the Reserves had familiarized me with military conditions, but flying to England in the belly of a Flying Fortress bomber was an education.

I was staggered by the beautiful complexity of these great skyborne fortresses—the nerve-endings of the dials, switches, knobs, handles, and warning lamps leading to the great vitals and muscles of these planes. There were eight or ten Forts at the army base where we took off, dark, squat and rooted to the ground by their tail wheels.

As I boarded, our pilot was supervising the prop-turning that would awaken this huge creature whose twin-lobed brain consisted of our pilot and co-pilot. Our lives were in their hands, but none of the military men climbing aboard seemed concerned as we squeezed together on the long hard benches.

I heard the whine of the inertia starter in the wing. A burst of blue smoke swept out into the prop wash as the number one engine caught. I heard the roar and felt the shaking of all that energy. Soon all four engines were going. As we taxied down the runway the great push of acceleration rattled my shoulder blades, spine and tailbone. We gathered speed, the tail lifted, and we were on our way to England.

It was a rough crossing. In spite of our flight suits, the minus fifty-five degree temperature at flying altitude was chilling. I'd shaved hurriedly that morning and my beard itched furiously inside my oxygen mask. So much for unpressurized cargo cabins. There was no possibility of conversing in the roar of engines.

The tight fit of the mask and the chafe of my beard almost made me want to jump. So to keep from going crazy in my frozen isolation, I reviewed my assignment. What with all the remaining headaches to be dealt with on the Burma line, I was not really prepared. But that didn't worry me. I knew all the key people in England. Over the past few years they'd each visited stateside pipelines to learn what they could to satisfy England's growing need for improved supply lines. I'd been their guide, often hopping around the country from one pipeline to another to show them solutions we'd developed for different climates and terrains. They were a dedicated and efficient bunch, and learned quickly. We had worked well together on these jaunts, and I expected to find their newly-learned techniques already implemented. All I had to do was check it out. After China, with its language and culture shocks, this one looked like a piece of cake.

Britain's pipelines were already in place. The main arteries, from Glasgow to Edinburg in Scotland, and from Bristol through London and on to Dover in England, had served the general peacetime business of the oil companies for years. Now British Petroleum, Royal Dutch Shell, and Shell Transport and Trading had begun adding to these lines and improving them for the war effort.

Two main things needed to be done. The first, getting fuel into Britain, required making connections from the Scottish coast to England. These had to be set up to ensure minimum damage from German attack. The English Channel had become too dangerous, and tankers could no longer be brought in there.

The second effort was to get down to the Channel from the existing southern line. This meant stringing portable pipelines, south from the east-west line buried under the Thames river, to seaports along the Channel. These ports would be the embarkation points for attacks on the German front in France.

My job was to evaluate the work being done and recommend improvements. The existing equipment was in good condition and reasonably up to date. The question was, were pipelines as ready as possible to move huge quantities of fuel, and had the new portable lines been placed in the best locations to minimize loss of tankers from torpedoes and bombings?

The choice of an American for this evaluation had been politically astute. The English needed somebody who would be objective, recommend changes, and leave the local crews to carry them out without having anyone to blame for them except outsiders.

I was used to making bogeyman decisions. They seemed far easier than riding in this fort, hungry, cold and entombed in my flight suit and mask. Anything on land seemed easier than this.

As we eased down, out of some low clouds the color of coal smoke, we saw England. After the strain, boredom and cold of those many hours over the ocean, it was not haycocks and hedgerows I was looking for but the black streaks of burned rubber from the wheels of other planes that had touched safely back to earth.

To the music of our thumping tail wheel on solid ground, we taxied behind a jeep with FOLLOW ME painted in big yellow letters on a high backboard.

It was raining as I stepped onto English soil. The water slanted under the plane's wing on a raw northeast wind. My impression through the deluge was of somber flatness and mud. Mud oozed up over the edge of the asphalt circle where we parked; mud filled in the tread of the jeep; a vast plain of mud stretched off toward a cluster of barely visible buildings. I hoisted my duffel bag and headed for the CO's office, comforting myself with the thought that at least military flights bypassed the long, boring formalities of customs and immigration.

Chapter XII

The Islanders

I checked in at Shell's luxurious Lensbury Club, about eighteen miles west of London on the Thames. During peacetime it had been used as a college for overseas managers. It now housed most of Shell's offices and provided living space for its executives. It would serve as both home and office for my stay, and was conveniently located near Hempstead, only a few miles from the underground tunnels where Britain's military offices had been relocated during the Blitz. Known as the Paddock, this War Cabinet citadel was fully equipped with offices, sleeping quarters, telephones and telegraph facilities.

The Blitz, the nightly attack on London by German bombers which had begun in late 1940, had ended. There had been no air raids for several weeks before I arrived, and the English seemed in a festive mood. London had been caught unprepared for these raids. Apart from the "tubes," there had been no really safe place. Very few cellars or basements had been able to withstand a direct attack. The whole mass of the London population had slept in their homes or in Anderson shelters for months, under enemy fire, taking their chances night after night. The Blitz had not seemed to have affected the Londoners' ability to carry on with whatever they were doing during the daytime, but now that it was over, the British were determined to return the favor. The Germans were in for some heavy bombing.

I went on an inspection tour of the Scottish and British pipelines. Things were in good shape, considering that English farm owners had inadvertently turned the lower pipeline into a target for German bombers.

Thinking to protect it, they had built an English stone fence over the buried fuel lines. Since these lines ran straight as a die for miles, the fence merely pinpointed the pipelines for air attacks. We set out to disguise them, first by painting the top layer of rocks a neutral color to blend with the countryside, then by surrounding them with people and livestock, installed as mini-farms along the oil companies' rights of way.

After the intrigues in China, the patriotism and cooperation of the people in England seemed miraculous. The British pipeline workers were meticulous about following instructions. Excellent craftsmen, they were able to produce almost anything I asked for, if given enough time.

After looking over what had been done to improve equipment at selected points, I arranged a labor force with London Shell, working out which oil companies would supply labor for each branch to be added. This went easily since the entire pipeline system was under one control, like a big freight yard.

The weather was no help on my inspection tour. One day it would be cold, wet and windy; the next would dawn beautifully but become cloudy by 8:30 A.M., rainy by noon, and windy in the afternoon. By evening, stoves would have been lit to chase away the afternoon chill. One day in June, nine separate fronts, some bearing hailstones, passed through. The next day, we had electrical disturbances. By the time the skies cleared in July, I was back in London.

On the Fourth of July, I attended a garden party given by the American ambassador. The party was held to honor Eisenhower's recent appointment as Supreme Commander over American military services in Britain. This was my chance to meet the American and English brass I'd be working with. I was still a Lend-Leaser, under the supervision of the War Department.

At the party, Eisenhower seemed impatient, even a little embarrassed by the abundance of rich foods and drink that ordinary Londoners had not enjoyed since their war had begun. Chain-smoking and occasionally flashing his brilliant smile, Ike strained to understand the thick accents of the indolently graceful ruling-class British. A tall, alert British aide hovered close by, helping him with whispers of information about members of Parliament or the British High Command who were present.

My introduction to Eisenhower was cursory, and I busied myself making contact with the British and American military supply men. In less than two hours I had met all the necessary people and made arrangements to see them the following week. At the top of my list was a meeting with Colonel John Clifford Hughes Lee ("Jesus Christ Himself" Lee), commanding general of the Service of Supply in the European Theater.

As I headed for the exit, Ike's aide introduced himself. "Known around here as the Gopher," he smiled as he shook my hand, "but our friend Donovan calls me Tom."

I'd met my British contact, for whom I'd be carrying communiques.

Within a few weeks it was apparent that Eisenhower planned a European attack. More and more American forces were landing on British shores. Portable pipelines had to be run to five destinations along the English Channel. Tensions were rising between the civilian and military populations. Our boys saw themselves as crusaders come to save the British and they were expecting to be treated like heroes. The Brits, who had been fighting alone for two years, resented the added discomforts and deprivations entailed by the arrival of the Americans.

Americans complained of the cold houses, warm beer, tasteless food, poor telephones and plumbing, and had trouble with the unfamiliar money and driving on the left side of the road. They regarded the upper-classes as snobs and lower-classes as slobs; and they felt contempt for earlier British military failures. They were suspicious of British imperialism.

The British resented the Americans' late entry into the war and their assumption that they were going to win it soon. They abhorred the racial discrimination they observed among the American military personnel. They had a saying that American servicemen were "overpaid, oversexed, overfed, and over here." The Americans had a reply: the British were "underpaid, undersexed, and under Eisenhower."

Ike's first job was to combine the two camps into a single cooperative unit, aimed at the enemy . . . not an easy job. I understood how our boys felt, but who could fault the British? For months, they had been attacked nightly by an average of 200 German bombers. These bombings had killed 44,000 civilians and wounded 50,000. In London alone, 375,000 had been left homeless.

One Sunday in July, during the banking holiday, I took a walk through London. The city seemed like a run-down clock—hushed, vacant, purposeless. The only sign of life was the pervasive cooking smell of cauliflower and brussels sprouts. The crumbling buildings looked like badly decayed teeth with giant cavities.

Some bomb-ripped buildings looked like stage sets, wallpapered rooms exposed to the world. A lavatory chain dangling in air over a floor that had fallen away, or a staircase that led to an empty sky, would lead the mind groping for a handle. Bomb sites were piled with rocky debris and the paving stones of porches that led crookedly to nowhere. Gardens were in heaps of masonry, with only a few flowers or rare plants sprouting from out of the rubble.

Over all this devastation, the washed-blue skies of summer brought hope. The sooty trunks of London trees were dressed with the delicate green of young foliage. A flower girl called out to me, "Cheer up, mate, ya' ain't dead 'til yer buried."

I decided to ride around. Mounting the stairs of a shiny red double-decker bus, I felt as if I were seeing London for the first time. It looked beleaguered. Vast stretches of barbed wire remained, along with barriers and pillboxes with menacing black gun slits. Weeds invaded boarded-up houses and the rotting burlap of sandbag walls. There were signs pointing to refuge in the tubes, whose stations were still lined with triple-decker bunks.

I got off the bus and had a cigarette in Regency Park, where people sat among the roses. I made change for a man in working clothes who asked, "Got the change of a quarter, mate?" and then sat on a park bench next to a distinguished-looking man. As we chatted for an hour I learned a few things about the character of the English.

"Adversity suits the British," he told me. "They like their roads to be crooked, their trains slow, their phone lines crossed. In peace they are in permanent resistance to authority. In war, they breathe a sigh of relief at getting back to normal. In solitude they feel happiest. They go back to being islanders."

Chapter XIII

Pumping Oil

The Allies attacked the German front in Dieppe, France. At the time I was back in the States. The attack had been a military failure but, as Donovan confided to me, other things of greater importance had been gained. I sat in my New York office for months, fuming over the supply problems that had added to the military failure. The supply problems seemed to be from poor usage of the pipeline rather than poor fuel delivery.

The oil people in England and New York were complaining about all the change-orders made by Eisenhower's supply people on the pipelines to the Channel. The military became our biggest problem. From the beginning they kept asking, "Is that the best you can do? How soon will it be done?"

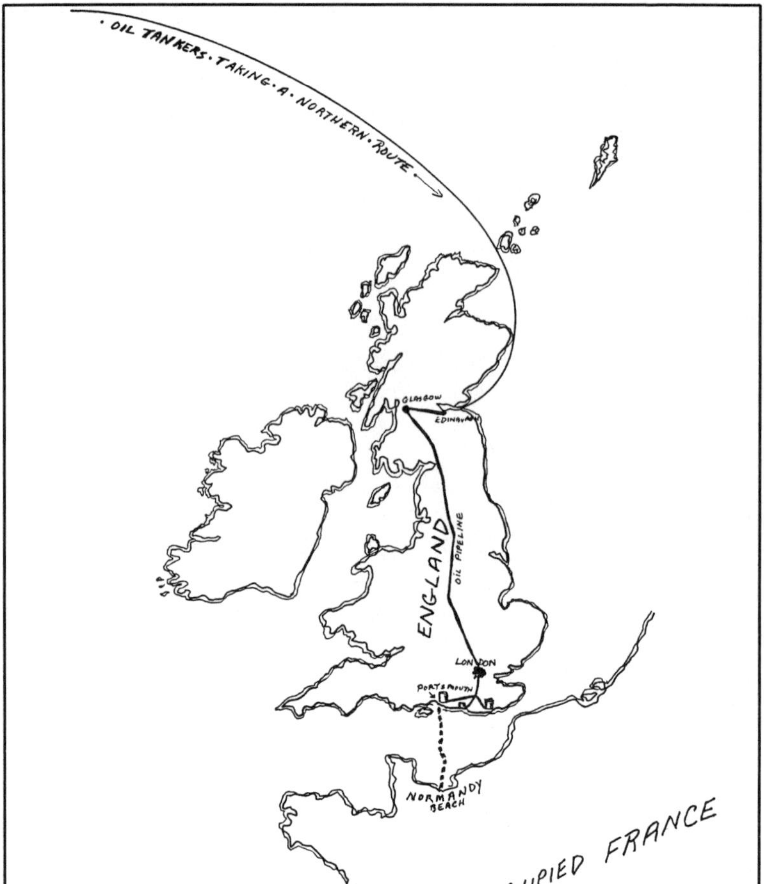

· OIL TANKERS · TAKING · A · NORTHERN · ROUTE ·
GLASGOW
ENGLAND
OIL PIPELINE
LONDON
PORTSMOUTH
NORMANDY
BEACH
NAZI OCCUPIED FRANCE

Their single focus was to get 25,000 barrels of oil pumped through the pipes in twenty-four hours. They weren't concerned with the build-up time needed to make sure that mixed products could flow through a single line without contamination. They expected us to deliver octane and aviation gasoline, several kinds of fuel oils and solvents, and pure alkaline, the synthetic base for 100-octane fuel, which enabled American planes to fly faster and higher and to hit harder than planes the Nazis or Japanese had developed.

Military men didn't understand the problems of running and scheduling a pipeline, or the electrical, mechanical and equipment preparations needed whenever supply requirements changed.

Back in England, I had oil men teaching military personnel how to operate portable pipelines without inviting attack or blowing up the lines. We taught them to check meter and gauge readings, reset connectors, change lines, and work pressure controls. We made them practice constantly and we ran a lot of safety drills. The changes, instituted by the brass, had to be handled on a different level. I passed the word to Donovan, and he came through for me.

By late 1942, plans had been laid for a final assault on the Axis. The highest level oil people had met with the War Department and agreed that the oil companies would meet military fuel goals, if the military would finalize their pipeline location requests.

To pin down where connections would have to be added to the system, I went to see Colonel Lee at his underground War Department offices. Orders were already coming through for deliveries to storage tanks between Portsmouth and Newhaven, and there was no more time for tinkering around with that section of the coast. The exact timing of the planned D-Day invasion was a tight secret, but we all knew the importance of these fuel supplies.

Winston Churchill walked into the room while I was in the office explaining that any further changes would cause serious delays in delivery. Colonel Lee introduced me and explained why I was there. Churchill nodded and said to us, "It is good to see all you people working closely together. Carry on." As he and the RAF officer accompanying him walked on, Lee and I, perhaps inspired by Churchill's expression of confidence, came to quick agreement.

That night I got a phone call from Tom Wainwright, the Eisenhower aide who was my contact for Donovan. He sounded excited. "Good to talk to you again, Jack. It's been a while. Are you free this evening? How about meeting me for a quick one?"

Our phone conversations were always light and casual. One never knew who might be listening in. "Sure thing, Tom. I could use a breather. Usual place?"

"Yes. Nine o'clock?"

I caught a jeep into London. Thanking the driver, I asked to be let out a few blocks from my destination and threaded my way through the rubble of darkened streets. Mostly by touch, I located the door knocker of the basement-level bar turned officer's club. I knocked, and a military police shone a blue penlight through the security hatch in the door. I held up my identification and the door opened.

It took a few moments to adjust to the subdued lighting. Enveloped by the warm smoky air of the place, I relaxed as the sultry, sexy sounds of Ella Fitzgerald came in from the phonograph by the bar. The place was comforting. The laughing young barmaids in their bright skirts were a welcome relief from the dowdy grey matrons I was used to seeing at social gatherings. They were excellent hostesses though not strong on wit and charm. I gave myself a few moments to enjoy the barmaids and the atmosphere, then looked around.

Tom was waiting at a corner table. He looked tired but maintained his crisp military bearing. We ordered drinks and went through the formality of exchanging a few pleasantries. But I could sense his impatience with our charade. "What's up?" I asked.

"I understand you and Lee have pinned down the portable pipeline locations," he said softly, leaning over the table toward me.

"I'd hate to try to keep a secret in this town," I chuckled. "But yes, we have. Why?"

"Does that mean you'll be heading back to the States?"

"It does. Tomorrow, in fact."

"Good. I have something of utmost importance for our friend. Can you go there directly?"

"Sure. I can arrange to land in Washington, then catch the shuttle to New York. Be glad to do it," I said, still grateful for Donovan's help with the military.

Tom paid for our drinks and, as we climbed the darkened steps to street-level, he shoved into my jacket pocket a package of Chesterfields. Inside was a micro-dot message for Donovan.

Chapter XIV

Normandy and V.E.

For the next two years I shuttled back and forth, carrying messages and acting as watchdog for the New York and London offices. It was my job to see that everybody connected with the pipeline operation was doing his job correctly. Many of our supply problems eased when American troops swept across the Mediterranean to Sicily.

In May 1943, when the North African campaign ended successfully, larger supplies of oil began to flow to England in preparation for the counter-attack on Europe.

At home, a minimum forty-eight-hour work week was proclaimed by President Roosevelt for all essential war industries in labor-scarce areas. Salvage drives were going strong. Americans collected tin cans, fat, wastepaper, and tons of iron and steel scrap for essential industries. Zoot suits and jive flourished. Despite shoe rationing, the jitterbug became the national folk dance. Soft-cover copies of *One World* and *God Is My Co-Pilot* appeared in the arms of women in slim skirts, bare midriffs and large hats.

When I returned to London in late May 1944, Bob Hope was touring the Pacific. GI's in both theaters were enjoying Bill Mauldin's Willie and Joe, those stoic cartoon combat soldiers ("Just gimme th' aspirin. I already got a Purple Heart.") and resonated to George Baker's "Sad Sack," that unlucky, confused, unkempt but well-meaning GI who was forever in trouble. Military morale seemed high, and I could smell something big in the air.

Nonetheless, Britain was in a state of subdued fear. Everybody knew the Allies would soon cross the Channel, but only a few at the top knew when. The French coast, referred to as the Far Shore, had turned into a mysteriously menacing place.

German propagandists had planted the idea that their *Atlantikwall* was invincible, and that they would set the Channel afire when the Allied fleets appeared. They spun an image of terrifying military, technological and scientific power: armies of close to two million, equipped with immense tanks and defended in the air by jet fighters; bombers said to be equipped with rocket-controlled glider bombs and missiles; an armory of terrible weapons including death rays, bacteriological bombs and radioactive land mines; fleets of hydrogen peroxide submarines; and even nuclear weapons. In the War Department it was whispered that Churchill was having nightmares in which the Channel became clogged with corpses.

Tension grew. The stakes were high and the atmosphere was more charged than ever. The military and pipeline people were uneasy, knowing all their chips were bet on the upcoming invasion. Even people who knew nothing of the difficulties of preparation felt the invasion would fall short of their expectations. At worst, even if the German propaganda were only half-true, it seemed possible the invasion could turn into the most ghastly nightmare of the entire war.

Our Western Naval Task Force assembled at Plymouth Sound. A few miles down the English Coast at Belfast Lough, the heavy bombardment group of Task Force 87 was preparing to land the U.S. 36th Division at the base of the Cherbourg peninsula in France. Force Utah steamed south by east off the western approaches to England.

On June 1, I made a final inspection tour of the coastal pipelines, and spent the next five days answering the growing concerns of Eisenhower's staff. I was in direct contact with field commanders, assuring them that all was in place. They were especially concerned about the portable pipelines that had been laid, pipe by pipe by stealth of night, across the Channel to within a few feet of the Normandy beaches.

On June 4, Force Utah arrived off Falmouth, a small port in Cornwall at the entrance to the Channel. From London to the Channel coast, a grim and steady determination pervaded. Nobody slept for more than an hour or two at a time. In the small hours of June 6, as I rechecked reports from oil tankers and pipeline branches, Bob Smith burst into my office shouting, "It's happening! The attack has begun!"

Smith, a Shell T&T man with an MI-6 rating, yanked me out of my chair and ferried me to War Department headquarters. As I checked incoming reports posted on the vast situation map that covered an entire wall, my heart was in my throat. The map was splotched with measle-like dots that represented flak areas. Toy airplanes and parachutes decorated the map, showing current positions of both Axis airfields and our attack forces. Lines and circles marking the locations of radar channels criss-crossed the map, and gold threads marked the portable pipelines stretched from the main line down to the Channel. The Channel coastline was not visible on the map, buried under a host of toy Allied ships and destroyers.

There were more gold lines than actual pipelines. Working with the military, we had set up several decoy lines, hoping they would draw enemy fire away from the functioning supply lines.

I lit another cigarette, although my throat was raw from days of chain-smoking. I had managed to comfort the field commanders, but wasn't as confident as I had sounded.

I knew all the right equipment had reached England, but many things could have gone wrong with the installations. We'd been unable to do much testing, afraid the activity would draw enemy fire. I worried that somehow the Germans might trace back to one of our pumping stations with small grenades and raise hell with the entire system. But everything seemed to be working. It was working!

Morale was high, the fuel was getting to the troops, and the control room began to feel like the two-yard line of the football field of a college stadium.

Late in the day, news of successful paratroop drops behind enemy lines had been followed by reports of victorious bombing attacks and the triumphant landing of ground troops and equipment; I stayed glued to the incoming reports. Now that the beaches of Normandy and Provence had been secured, the main supply lines to our forward-moving ground forces were the ones running across the bottom of the Channel.

Things could still go wrong. A sudden load on the supply system could knock out the pumping system near the front end. This equipment had been tested under enemy fire only once, briefly, in Burma. What if a motor overheated or a pump ceased whirring? Would the warning buzzers, signaling the crews to bypass a pump or an entire station, be heard over the roar of battle? What if the lines developed thumps, hydraulic pressure on the valves that block the flow, as we switched the types of fuel through the lines? We'd put loops in the pipe to reduce this but if they proved insufficient, warping would cause loss of pressure in the lines and valves.

But our men had raised the underwater pipelines and unstopped them. The fuel was flowing! Temporary headquarters were under construction right there on the Normandy beach, near the pipeline. I let out a long-held whoop and collapsed. My job was done for the time being.

The next day, with nothing to do, I walked the streets. London looked worse for wear but glad to be alive. American troops and British workers hugged and danced in the streets. They had become a team, and their team was winning. The Germans' pilotless V-1 bombs, which began dropping on London a week later, could not ruin this team spirit. Even when V-2 supersonic rockets fell in September, the Allied spirit was undeterred.

♦ ♦ ♦

That July, back at Shell in New York, we celebrated the dedication of the world's longest oil pipeline. The Big Inch, as it was called, stretched 1,254 miles from Texas to Pennsylvania, crossing 30 rivers and 200 streams. This cross-country pipeline had been started because enemy submarines had been sighted off the U.S. coasts, and threatened to destroy tankers and precious oil. That threat no longer existed. Now, with optimism running high about a victorious end to the war, the oil industry was thinking about peacetime uses for the Big Inch. Post-war needs became the focus.

In August, American armored units were moving rapidly across France. By year's end Allied forces were poised on Germany's eastern and western borders, preparing for the final assault.

On December 16, the Germans made their last major offensive, the Battle of the Bulge. Bad weather hampered Allied air and supply operations, but on December 23, the skies cleared. German armored columns were halted for lack of fuel and we destroyed them. Just like I'd said to Bousfield back in Hong Kong, whoever gets the oil wins. It was all over but the celebrating.

Chapter XV

Peace and Prosperity

Virginia and the kids watched the ticker-tape victory parades with me from the Shell offices high in the RCA building. Now that the war was over, Shell would soon be making good its pre-war promise to put me back into Marketing. I was using some built-up vacation time to scout out the best school districts and residential areas around New York City.

I had mixed feelings about the festivities on the streets below. Like everyone else, I had rejoiced over Germany's unconditional surrender in May; ending nearly six years of war. Although the Japanese surrender in August was welcomed, I was troubled about the destruction of Hiroshima and Nagasaki. The atom bomb had effectively ended isolation for the United States and put us in the role of peacemaker in a disordered world.

At home we faced spiraling inflation. House hunting was going to be tough in the midst of an acute housing shortage. Collective bargaining had been restored and labor disputes that had remained pent up during the war now burst into the open. Soviet expansion became a new threat to world security, and poverty-stricken Europe was looking to us for aid.

I kept these concerns to myself as we enjoyed the pleasures of New York. The kids were enchanted with *Carousel*, Rogers and Hammerstein's musical at the Majestic Theater. They gobbled hamburgers for lunches and dinners while Virginia and I enjoyed steaks. The three of them did a lot of sight-seeing and most of the house-hunting while I conferred with Shell officials about my new duties.

A lot had changed since 1942. World War II marked the end of an era for Royal Dutch Shell. The siege of Europe, which had dispersed head office personnel out of the Hague for five years, resulted in a wartime skeleton crew operating out of London and Curacao. Shell's Far Eastern and European production and refining capacities had been knocked out. Marketing had problems. The changeover from war production to consumer goods would take time. The free trading and transport pooling our government had allowed the oil companies during the war might soon be withdrawn.

The good news was that we stood a chance to outrace our biggest competitor, Standard Oil of New Jersey (Esso, which later became Exxon). In 1942 the U.S. Justice Department had accused Standard Oil of making restrictive agreements with I.G. Farben, a German chemical combine. To avoid criminal charges, Standard had signed a decree declaring that neither they nor their controlled affiliates would enter cartel agreements. With Standard Oil out, the As-Is agreement (which had cartelized the oil industry in 1929 based on each company's market share at the time) was finished and it was a new ballgame for the oil industry.

I spent the rest of the year in Zionsville, catching up on pipeline developments done in my absence. We had been experimenting with pumping various grades of fuel and continuing the pipeline additions begun during the war when our 3,000 tankers had been drafted to supply our forces. Now the Atlantic and Pacific Northwest states had their oil piped to them.

Our tank farms were brimming with gasolines, fuel oils, napthas, and two dozen other petroleum derivatives. The Atlantic seaboard was clamoring for supplies and we sometimes piped as many as seventeen different products simultaneously through the lines.

I also made frequent trips to New York, being groomed for my upcoming job as Manager of the Engineering Marketing Division—an area I hadn't worked in since I'd left St. Louis in 1939 to run the East Products Pipeline for the war effort.

During the Christmas school break we piled into our 1942 Buick Century and moved from Zionsville to Montclair, New Jersey, where Virginia had found us a big old fixer-upper of a house. Originally a church rectory, the house had not been maintained nor improved since 1929, when the latest owners had lost their money in the stock market crash.

On our way, we visited Virginia's parents in Illinois and we spent several days with my parents in Webster Groves, just outside St. Louis. I hadn't been home often since the Depression. Then, St. Louis had been so smoky and coal-polluted that headlights were needed even for daytime driving. I'd been glad to get away from this depressing place, but now the sight of our plain white-frame house sparkling in the afternoon sunshine brought back happier memories.

Things had changed. The fire station was still just down the street, where I'd chased the fire trucks with my kid brothers tagging behind. The horse-drawn fire wagons had been replaced by engine-powered models. The grammar school was still across the street, and I could see the spire of the Episcopal church, where I'd supplemented my allowance with the dollar I earned by singing in the choir every Sunday. When my voice changed I took a job jerking sodas at the drug store. Was the drug store still there?

I wondered if the Women's Club building still stood and if Mother still taught kindergarten there. Did the Club still run dances for the young people, as it had in my high school days? Was the Texaco station I'd worked at still doing business? If it was, it had surely changed a lot since I'd worked there. Back then, full service had been the rule, and that included filling the radiators and hand pumping air into the tires. The single air tank in the station had been used for lube jobs only.

There was new shrubbery in front of our house. I was happy to see that the sleeping porch we added when I was thirteen was holding up. John and Katie were going to love that. I loved it too, though at the time we built it I had resented spending the money.

I'd had a job driving new cars from the St. Louis factory to the Webster Groves dealership at a dollar a car. It took a thirty-minute trolley ride to get to the factory, so I'd only been able to deliver one or two in the afternoons on school days, or up to three on Saturday mornings. I had loved that job. It was exciting and had made me rich while it lasted. I'd been driving since I was ten and had a perfect driving record. The road hazards of the day (inexperienced drivers and absent-minded pedestrians, unused to cars, who stood or walked in the streets) were no problem for me. I was respected for my driving skill and proud of never having an accident. I'm still proud of that.

Mother came out to greet us and to hug her grandchildren. She had seen little of them during the war. Dad, still practicing law (as had his father and grandfather), was busy with a client. We didn't see him until the evening meal, when he strode into the dining room and vigorously shook my hand—the extent of his demonstrativeness—before greeting the rest of the family.

Over dinner, Dad entertained John and Katie with tales of my childhood. Katie giggled over the story of my setting fire to a neighbor's hedge because he wouldn't let us kids cut through his yard on our way to school. Dad showed Katie the razor strap he'd used on me after that caper, and Katie said she guessed her mother's paddle wasn't so bad after all. John was impressed that I had done the driving on a family vacation to Niagara Falls in my early teens.

The kids played outside while we had coffee. I was telling Dad about my new job in New York when Mother said, "Tell me, Son, have you ever regretted going to Washington University instead of Purdue?"

I hadn't thought about Purdue for years and the question startled me. I'd been excited about my athletic scholarship to Purdue, but when Mother contracted pneumonia a week before I was to leave, I'd decided on Washington University instead. In those days pneumonia was often fatal, and attending college only a half-hour away meant I could live at home, at least until her crisis passed. It also meant the end of my sports career since the practice and travel time for team members would keep me away too much.

I smiled at Mother, who sat waiting intently for my answer. "Never, Mother. If I hadn't gone to Washington University I wouldn't have gotten into ROTC, and that's had a lot to do with where I am today."

While Virginia called the kids in from the gathering darkness and got them ready for bed, I told Mother and Dad about my work with the military in China and England. Mother went to bed happy that night, relieved that her illness had not disrupted my life.

We took our time on the rest of the trip, sight-seeing along the way. This was our first real family vacation since before the war. I knew time would be scarce in New York. Managing the Marketing Engineering Division was an exciting prospect, but Shell's goal of doubling sales was going to keep me running.

Oil supply had been the big problem during the war but now, with government continuing to allow oil trading between companies "for the good of the economy," we had to cultivate customers competitively. Shell had begun geographic management mergers and there would be a lot of retraining and regrouping in my division.

As these business concerns went through my head, we reached Montclair and turned into the tree-lined streets of Upper Montclair Avenue.

"This is *great*, Dad!" Katie shouted, looking at the handsome houses we were passing. I pulled to a stop before our new home. Her enthusiasm vanished. "That's the ugliest house I've ever seen," she pouted.

The place was gloomy, with its ancient, peeling paint and neglected yard. But it was big, structurally sound and the best house available in the areas where we wanted to live.

"I know it's not much to look at now, Katie, but wait until John and I get our hands on it. We'll have it looking good in no time," I assured her.

"No time" turned out to be longer than any of us expected. John and I replaced the old floor in the kitchen and butler's pantry, updated the wiring, installed a new entrance, and had the place replastered and reroofed. In six weeks we moved out of our temporary rental into our new home. Our first meal, a steak dinner, was cooked in the kitchen fireplace. John, Virginia and I were celebrating. Katie was still not happy.

Going from a semi-rural midwestern setting to a modern school in an international bedroom community was a big change for the kids. Katie missed her old friends and John suffered being demoted from top dog to just one of the pack. The renovations were moving along quickly, but our house was still the most decrepit in town and that didn't help the kids' confidence.

At first Katie and John kept close to home. Virginia and I taught them to play bridge and signed them up for golf lessons at the local club. Soon they were begging off from family activities to join their friends: for John, softball, soccer and tennis; for Katie, band and Girl Scouts. John, an enthusiastic helper on the initial house remodeling, got scarcer and scarcer.

Our four-story, fourteen-room house restoration became an ongoing one-man project, fitted in on the weekends I was home.

By now, I was deeply involved in Marketing Engineering, preparing the division to increase sales. We had to shuffle a lot of personnel as retirees were replaced by new employees and those returning from military service. Returning employees were guaranteed their old jobs, but their jobs had changed. These men needed reorientation and a chance to catch up on new developments.

Marketing Engineering included Shell service stations, water terminals, the delivery vehicles fleet, aviation fuel facilities, tankers, and working with auto designers to devise appropriate grades of fuel, engine oils and lubricants. My three biggest jobs were revamping Shell's network of service stations, supervising inter-plane pipeline services, and reorganizing Shell's plants and personnel.

Older guys started retiring left and right once the war was over. I began conducting schools day and night to teach new men their jobs. Their job pressures were small, since restrictions were still in effect on gas and furnace fuels. I had them all attending classes after their work shifts.

Doubling sales to service stations and commercial consumers, called for imaginative changes. Bright new service stations replaced older ones and were designed to reflect regional taste.

As interstate highways began to crisscross the country, ranch-style stations spread in the West, while a Cape Cod style identified Shell in the northeast. We tore down ten old stations and replaced them with four larger ones, well-located, which did more business than the original ten. I put together new guidelines and ran monthly contests for our sales staff.

I did a lot of short-hop travelling from 1946 to 1951, working with high school principals and company-owned service stations to create training programs for young men interested in becoming service station operators. Then, there were meetings to attend and committee work to be done for API, the American Petroleum Institute.

On airline pipeline services, I found myself working with airport authorities, airlines, labor unions and plane manufacturers. And there was always a lot to do for the many employees under me. Besides the periodic salary and performance evaluations. I had to see to their personal needs, such as making sure employee transfers coincided with their children's school schedules and helping them get a good deal on new housing. This took time, and budget expenditures needing my approval kept me even busier.

Marketing Engineering was anything but a routine job, and was further complicated by fragment reorganizations on the executive level. In 1949, I assumed the newly-created position of Assistant Operations Manager, and two months later the Operations Manager and I were switched. Suddenly, I was left with handling department personnel whose loyalties were with their old boss whose job I had taken over. I had to fire a few men who resented my authority, making it more painful for me to oversee Shell's physical operations. I had to go on faith that Jordan, my boss, knew what he was doing in switching me with the former manager. My job was to execute his orders.

Life in our new home went on. I rebuilt the steam radiator system and extended the driveway to the back entrance. Virginia had her hands full with the kids, now in high school. Katie was busy as a Girl

Scouts Cadet and with her piano and school band commitments. John had caught the car craze.

He bought a 1941 Crosley truck, which he used for his many businesses such as landscaping, building driveways, and delivering products of the print shop he and his friends operated in our basement. He took the truck apart in our basement. Virginia put up with the noise and car parts all over the house, but blew up when he painted it. A hideous green film drifted up the vents and settled over all the furniture on the first floor.

Except for avoiding piano practice, the only problem we had with Katie was over boys. Her tastes differed from ours. We settled this over the dining room table, where family conferences had become a new ritual.

Our dining room talk was wide-ranging. The kids often started involved conversations, sometimes about personal or family business but often about current affairs. I never cared what position they took on world affairs, but would hammer at them for a logical defense until they either defended their positions or backed down.

I think the many sophisticated people they now came into contact with enlarged the kids' scope. People from all over the world lived in Montclair and belonged to our country club. Virginia and I often entertained executives from Shell and other oil companies, plus executives from the many types of companies we did business with.

We often had London Shell people as house guests, giving them a chance to see how their American cousins lived. All of them liked Katie and John and went out of their way to include them in our conversations. The house guest that most inspired the kids was Bousfield.

Chapter XVI

Diplomats and Kings

I hadn't seen Bousfield since China in 1942. His job, which included parts of the Middle East and Africa as well as his Far East territory, had grown in importance by the time he spent two weeks with us in the spring of 1950.

The kids adored him, finding his style "mysterious" (Katie) and "hot" (John). Virginia, too, sparkled in the presence of this charming man. I lost a lot of sleep having late night talks with him.

Bousfield had a lot of discriminatory powers in his realm. As second in command to E. B. Mayne, his power extended over the hundred or so companies under Mayne's jurisdiction when Mayne was away from his desk. While Bousfield's powers did not extend to independent affiliates such as U.S. Shell, he spent a lot of time in New York because of the Group's many subsidiaries located there.

Several other major oil companies had headquarters in New York and Bousfield was often involved in negotiations with these companies regarding oil interests in foreign countries. The Group was doing business in nearly every major language and in more than fifty currencies, and needed to adapt to innumerable and often contradictory laws and business customs. International consortia and other joint projects were always complicated, so conversations with Bousfield were always an education.

One evening, just after congratulating me on my recent promotion to Assistant Vice President of Marketing for the U.S., Bousfield said, "Of course, it's too bad you're stuck in the marketing end of the business."

"Why is that?" I tried not to sound insulted. Since 1945, marketing had done marvels and I was proud of my part in that.

Through the American Petroleum Institute and our Washington lobby, we had achieved remarkable supply controls and some hefty tax privileges. U.S. domestic oil production was now held to predicted demand levels. Internationally, overall oil production was limited to a predetermined growth rate. These production controls allowed us to make several price increases in the States, and these were immediately transmitted throughout the world.

It seemed to me that U.S. marketing was the fair-haired boy of the industry, responsible for growing profits around the world. Where the hell did Bousfield get off, talking about marketing as if it were a step-child? It seemed to me that marketing was especially important to the Shell Group, which had always been better at selling oil than at finding it.

"The industry has benefited tremendously by being able to control world production levels," Bousfield said. I felt less ruffled with this admission.

"But that could boomerang on us now," he continued. "Post-war demands have increased faster than anybody thought. A lot of that growth has been in the Eastern Hemisphere . . . unprecedented industrial expansion going on there, where about half the Group's worldwide sales are made. We do about a quarter of our business there now, and it's going up. You know what that means, don't you?"

I didn't. I had kept up with the people I'd met in China but I didn't know much about the Middle East or Africa, beyond the fact that we were able to buy unlimited cheap oil from those areas through London Shell. Although more than half of our fuel now came from domestic oil fields, U.S. oil reserves were considered endangered and we were buying more and more foreign oil. I only knew that we had been moving farther and farther from home for our purchases.

Mexico had nationalized its oil fields in 1938, disenfranchising the major oil companies who had explored and developed them. The Group, like the other majors, had turned to Venezuela that same year. All it had required to get Venezuela to agree to a forty-year contract was higher royalties. Venezuela had provided the majority of oil used for the war in Europe, and was the world's chief exporter during the war.

But in 1948, Venezuela demanded fifty percent of the profits from its oil fields. Esso, Gulf, and the Shell Group (the companies most involved) had accepted that, reasoning that it would make foreign-owned corporations less vulnerable to nationalist attacks. Then the companies stepped up exploration and development and soon doubled their output of oil, earning as much as before the profit split. They had also invested heavily in Saudi Arabia, where the profits were even larger.

"So why is Eastern growth a problem? Can't we just increase production in Venezuela and the Middle East to meet higher demands?" I reasoned that all it would take would be taking off the cap on the production limits the major oil companies had established.

"The problem is that those antitrust moves by your government have put ideas into the heads of the rulers of the oil-producing countries. Now they all want a piece of the pie. Have you heard about what's going on in Saudi Arabia?"

"Only that the king is spending his new oil wealth like there's no tomorrow. Is there more?"

"Considerably more, and your government is right in the middle of it. It's very complicated." I smoked a cigarette while Bousfield thought for a few moments. "First," he continued, "there's the antitrust mess that's been brewing in the States since before the war."

"You mean the subpoenas and documents the Justice Department collected from the oil companies before the war. The ones they had to sit on by the State Department while we supplied the war effort?"

"Precisely. As you know, Justice is again breathing down the necks of the five American majors."

"Yes. I haven't kept up with the details, since the Group is outside U.S. jurisdiction."

"Yes. London has its advantages—which puts your U.S. Shell in a very fortunate position. You have the advantage of buying through London at the 'family' rate but, because you operate as a wholly independent company, your government cannot hold you responsible for any of the Group's corporate dealings."

I had to smile at this understatement. While the other U.S. majors were sweating out the Justice inquiries, Shell was concentrating on an unruffled expansion. "I'm sure it's no accident that we're set up as a distribution and marketing firm, with oil field ownership limited to the U.S.," I said to him. "That leaves us spotless, and makes you guys in London and the Hague the bad guys."

Bousfield frowned. "It is only your government that frowns on the oil industry's control. England and Europe are more realistic. Until Europeans discovered and developed sources around the world, oil was an unknown and useless commodity. It was out of reach to the ignorant and savage peoples who squatted above it. We Brits are proud to have brought the Middle East and Africa out of the Dark Ages and into the twentieth century by giving them a modern industry to export. And until recently, those countries were grateful to us for doing so."

"But you're going to see industry-wide changes as a result of these antitrust scandals. It's giving ideas to third world producers, not least of which is King Abdul Aziz who, as you noted, is spending faster than he's earning. He wants a fifty percent tax like Venezuela's."

That was interesting news. Until 1943, Britain had subsidized the King, to the tune of around $20 million. Then Roosevelt, who understood that wars are run on oil, declared Saudi vital to the defense of the United States. By the end of the war the U.S. was the dominant influence in Saudi, subsidized through American Lend-Lease.

To the outrage of British Petroleum and the other American majors, Texaco and SoCal (Aramco) were now sitting on the biggest and cheapest source of oil in the world.

"So your State Department, worried about the Communist danger in the Middle East, and Aramco, who has the Saudi concession, are working up a scheme. They're strong-arming Justice into lowering its antitrust charges. *And* they're putting together a package that will leave Aramco's earnings intact."

That wasn't going to increase Aramco's popularity in the industry, but I supposed its stockholders would cry all the way to the bank.

"It won't be finalized for a few months yet," Bousfield continued, "but what they've cooked up is a tax deal for oil. Aramco will go ahead and pay the King his fifty percent, and these additional payments will henceforth be regarded as constituting a foreign income tax. Isn't that beautifully simple?"

I thought about that for a minute before the light bulb went on. "Does this mean that, under the existing rules against double taxation, Aramco won't be taxed in the United States?"

"It does indeed. It boils down to a subsidy for Aramco in Saudi Arabia, paid for by U.S. taxpayers. My guess is that in its first year, this ruling is going to cost the U.S. Treasury about $50 million in taxes. What do you think about that?"

"Well, it's a way to give foreign aid to a strategically important kingdom without having to go through Congress."

"Exactly. Involving Congress would be particularly embarrassing since the United States is publicly committed to supporting the State of Israel, which is fighting the Arab nations for its survival. This little tax gambit allows your government to openly support Israel, necessary for the votes of the American Jewish bloc, and to also support Saudi Arabia—ever so quietly, through the oil companies—because America needs the oil. In London we're calling it the Golden Gimmick."

"I see. Aramco gives the money to Saudi instead of to the Treasury, and the State Department buys Saudi allegiance without a penny of it showing up as State's expenditure, right?"

"Precisely, Hall. Precisely."

"It seems to me that this will give the other American oil companies a big incentive to invest abroad. They'll all want in on this tax advantage, won't they?" I hoped I was following Bousfield's complicated tale of governments and oil.

"They will, and the Group is moving to counteract that. But it does something else, which pleases me immensely: it makes oil an important instrument of your country's diplomacy!

"British Petroleum, of course, with its government's half-ownership, has always had the ear of Parliament. That's what's made Britain such an antitrust haven. And now, you see, oil is becoming indispensable in this country, too. Marvelous, don't you think? Just like the old days!"

I saw what he meant. The State Department would run interference with Justice in order to use oil as an unofficial instrument of foreign policy. Oil had become important for national security, a way to hold down the spread of Communism. I wished I knew more about the Middle East situation.

Bousfield continued to chortle about oil's growing world power. But I wanted to know how this tied in with his remark about marketing.

"Quite simple, really. This tax thing, combined with the world's increasing demand for oil—did I mention that western Europe is expected to double its oil consumption in the next decade and the Eastern Hemisphere to increase consumption by eight percent a year? Well, the tax breaks and increased demand are likely to change the executive makeup of the oil companies."

It looked like he was finally getting to it. I settled back in my chair.

"A lot more energy will be put into exploration and production," Bousfield predicted. "The companies will be preoccupied with crude. That means we'll see fewer marketing men at the top, with more engineers and geologists dominating the boards. You're an ambitious man. You can see for yourself that your engineering skills may soon stand you in better stead than all this marketing expertise you've been acquiring."

I was getting angry again but that didn't slow Bousfield. He was really warming to his topic.

"I believe those service stations you've been littering the freeways with will soon be of little interest at the top. With more cars and all the other growing demands for oil, customers will be 'in the bag,' as you Americans say. There'll be customers aplenty for every drop of oil we produce. So, finding, refining and shipping it will take precedence over romancing customers."

"Fine," I said curtly, "but somebody's still got to get the oil to the customer, and it might as well be me. I'm good at it."

"You're good at a lot of things, my friend. Why not hitch yourself to the rising star and give yourself a chance at the board of directors? You're only forty-three. There's time. Or have I overestimated your ambitions?"

His suggestion troubled me, more from loyalty and absorption with my job than lack of ambition. I'd built a good team and good customer relations. Was he suggesting I abandon that, just as the years of reorganizing were beginning to pay off? Bousfield must have guessed my thoughts, for he interrupted them by saying, "I realize you have put heart and soul into your sales goals, and that is commendable. But you're too good a talent to be wasted. You've done what you set out to do. Staying would be resting on your laurels, and I think you'd soon get bored with that."

He was right, of course, and I was mollified by his praise. "What exactly are you suggesting I do?" I asked, all the stiffness gone from my voice.

"I suggest another overseas assignment. Are you game?"

Chapter XVII

The World of Oil

"Mr. Mayne is free now, gentlemen. Please follow me."

Dempster and I walked behind the very proper young Englishman, down a long wide corridor blanketed with dove-grey carpeting. We passed ornate portraits of the Dutch and English who had made the Group a major player in the world economy, with holdings in nearly every country, colony and principality outside of the Iron Curtain.

We were ushered into a huge sumptuous office. At the far end, behind a massive desk of gleaming walnut, sat E. B. Mayne, one of the seven directors of the whole show. Mayne, in charge of the Group's organizational planning, was a man of quiet forcefulness. His presence filled the cavernous office.

I could see George Dempster was as impressed as I was when I had first met Mayne in 1942, nearly a decade ago. Our meeting had been brief and to the point, concentrating on the wartime emergency that had brought me to England. There had been no further contact with Mayne beyond his letter of thanks and subsequent commendation to my boss on a job well done. That letter, added to the one from Eisenhower, were my proudest wartime possessions—the Lend-Leaser's equivalent of the Purple Heart, I thought.

Now I was here again, on loan from Shell Oil Company to its lifesource, the Group. Dempster and I had been hand-picked by Bousfield for the job and we both considered it a high honor. And it was proof of the terrific marketing job we had done in the U.S.

Our destination was the Middle East and Africa. Our assignment was to bring the Group's organizations in those areas in marketing and operations up to U.S. standards. The reason was competition.

"Thank you, Geoffrey. You may leave us," Mayne said to our escort. Rising and extending his hand, he said, "Good to see you again, Hall. And this is Dempster, the other half of the team? How do you do."

"Please be seated, gentlemen, and welcome. I hope you and your wives are enjoying your stay in London. You will remain here for several weeks, being briefed on your assignments before proceeding on to Israel and points west. If there is anything we can do to make your stay here more pleasant, please notify Geoffrey. He'll see to all your wants. Now. Your assignment."

Mayne settled into his chair, giving us each a friendly but piercing look.

"You know what you are to do, of course, but I believe it is important for you to understand why we have found it necessary to take you away from your regular duties for six months."

"The marketing of oil outside the U.S. is more competitive today than it has been at any time since the price wars of the 1930's. We handled it then with an international oil cartel, but that came to an end with Jersey's consent decree in 1942. Without Jersey, the As-Is cartel became meaningless."

"However thanks to the merger of Royal Dutch and Shell Transport, we, the Group, are now completely international. Not only is the Group owned binationally—sixty percent by Royal Dutch of the Hague and forty percent by Shell Transport of London—but we operate through 500 subsidiaries and affiliates around the world. U.S. Shell, of course, being one of these.

"But we remain the world's second largest oil enterprise, still second to Jersey. Now, however, we are in a position to change that.

"The Group does half its business in the Eastern Hemisphere. Oil consumption is expected to double there in the next fifteen years. Jersey does most of its business in the Western Hemisphere, where consumption is not expected to double for twenty years. In words, gentlemen, the Group can grow faster than Jersey."

I couldn't suppress my grin. A quick glance at Dempster showed his pleasure in what we had just heard. Standard Oil of New Jersey (now known as Exxon), our biggest U.S. competitor, was about to eat our dust globally, and we were going to help bring that about!

"If we and Jersey both maintain our relative positions in the market, only a quarter of Jersey's sales will increase as fast as the Group's sales. Then, too, the Group's petrochemical business, already more than $400 million a year, could double by 1962. So you see, the Group could very well double its total volume by the late 1960's.

"This hasn't been accidental. We've invested at a higher rate and grown faster than Jersey since the war," Mayne continued, "more than the oil industry as a whole, in fact. We plan to lead the pack. But to double our volume we have to maintain our present market share, and the market keeps growing."

"Globally we're expanding our supply of crude, of course, especially in the Middle East, which lies in the middle of our expanding markets. With the world's largest tanker fleet, we can move oil efficiently anywhere we want. All we have to do is keep ahead of our ambitious competitors. That's where you come in."

I leaned forward, listening intently.

"We plan to build the refineries and marketing organizations needed to double sales. We need your American-style aggressiveness, your American concept of the steps of competitiveness that's worked so well in the States. We want you to look at the marketing in our

Consolidated Areas and work with the local managers to bring them up to U.S. standards. Can you do that for us?"

Fired by the challenge and both talking at once, we assured Mayne we could.

The next weeks were a blur of meetings with executives who administered the Consolidated Areas. Virginia and Eileen Dempster spent their days sight-seeing and shopping, making last-minute purchases for the tour of the Continent they would make while Dempster and I travelled through the Middle East and Africa.

Virginia was excited about this trip to England and Europe. She was doubly glad to be going with Eileen, who promised to be a pleasant companion. I was glad she would not be touring the Continent alone, and glad she would not be coming with me. The Middle East was politically volatile and Africa was promising to become so.

Virginia and Eileen joined Dempster and me for the daily cocktail parties and dinners arranged by the Consolidated Areas executives and their wives. London was seeing to it that we met all the right people, in business and government, and I ended each day adding notes to my growing dossier. I was learning a lot about both oil and politics in the Eastern Hemisphere.

It was sometimes difficult to separate the competitiveness of the oil industry from governmental competitiveness. At the higher levels, where men moved easily between their governmental and industrial interests, there was often no difference. This was made sharply clear to me on the evening of February 10.

The cocktail parties were beginning to pale, and I paid scant attention as I was introduced to yet another lord and member of Parliament. This one was also highly placed in British Petroleum, which was half-owned by the British government. The man showed signs of too much drink and promised to be a bore.

I was startled when he said to me, "You Americans beat us out of Saudi Arabia with your damned Lend-Lease bribes, but make no mistake, you're no more popular than Britain is. The Arabs aren't going to forgive either of us for recognizing Israel—you especially, after Roosevelt's promise to Saudi that your government would make no move on the Palestine problem without their consent.

When Truman recognized Israel within minutes of its declaration of statehood, the king felt betrayed."

"But surely he recognizes that a change in leadership often results in policy changes," I countered.

"He does not, and Arab memories are long, my friend. They were wary before, with your State Department favoring Saudi while your Executive supported Israel. But they were persuaded of your good intentions by all that money, backed by Roosevelt's pledge. That pledge was not honored, and the money will not be enough."

"No, Lend-Lease money won't hold the king's allegiance like Islam will, and your new Middle East oil domination may tumble, as ours has. But we still control Africa and mean to keep it. It's too late to stop the competitive oil explorations in Egypt, but south of there we're in charge."

I didn't fully understand the complex interactions between oil companies, their home governments, and the governments of oil-rich nations. But I was learning fast.

On February 11, 1951, Dempster and I boarded a plane for Cypress. We were on our way to Israel.

Chapter XVIII

The Middle East

Israel had been declared a state only three years earlier and had been engaged in constant skirmishes with her Arab neighbors ever since. The Arab world resented the United Nations' granting of land in that area to the Jewish people, who represented only a fraction of the territorial population, and they had a point.

The Arabs reasoned that they had not mistreated the Jews in Europe and therefore they should not be the ones burdened with reparations. They said, let the Jews have Germany, the home of their oppressors. Meanwhile, Palestinian refugees had been displaced by the State of Israel throughout the Middle East and into Africa. And Israel continued to fight with Egypt, Syria and Jordan over boundaries.

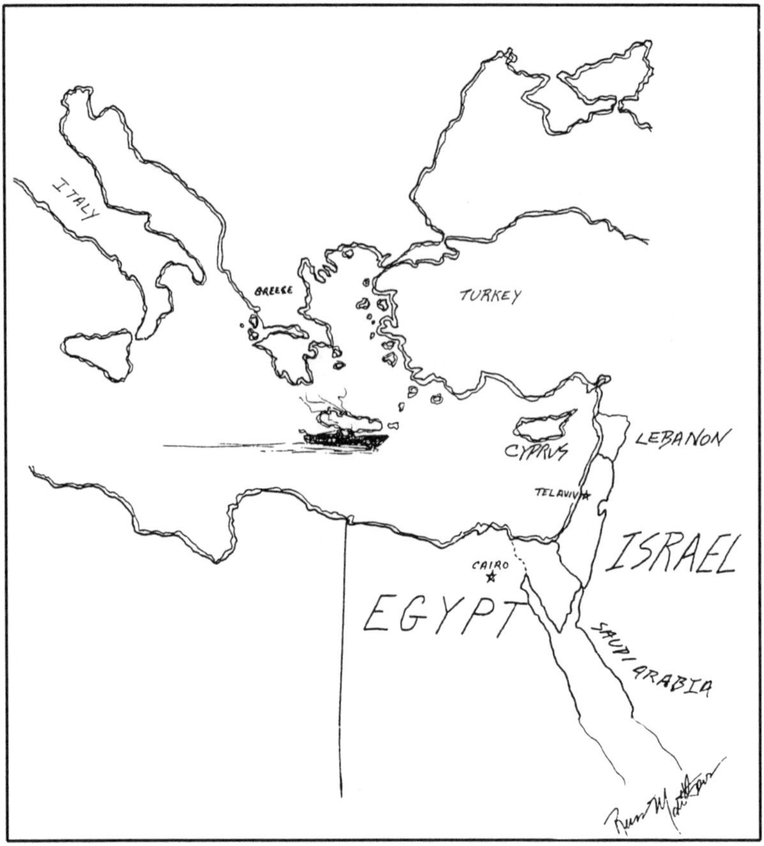

Our travel arrangements were complicated by Arab resentments. Though our business route went from Israel to neighboring Lebanon, Lebanon would not accept passports coming in through Israel. We were therefore flying into Haifa from Cyprus on a special visa, and would have to return to Cyprus to pick up our passports and enter Beirut via the Cyprus route.

While these political issues complicated our travel arrangements, they were not affecting the flow of oil. The consolidated area, known as The Shell Company of Egypt, Ltd., covered Lebanon, Syria, Jordan and Israel as well as Egypt. Its interests in Israel were restricted to pipelines and refineries. Iraqi crude oil was piped to the Israel refinery, from which it went across the Mediterranean Sea to European markets. Egyptian oil was shipped as finished product across the Suez Canal to supply dealers and retail outlets throughout the area.

The pipeline through Israel, Syria, Lebanon and Jordan continued an uninterrupted flow, operating under this single Consolidated Management Area, which largely ignored political problems and stuck to business. I only hoped oil's apparent political immunity would hold true throughout the trip.

Israel's exploding population was expected to continue swelling, but it didn't take long to determine that the Cairo office's proposed pipeline extension from Haifa to Tel Aviv was not a wise investment. I had looked it over at the request of Agnew, whom I had met in Cairo on my return from China. He had been of great help in facilitating my inspection, in 1942, of materials on the Cairo docks that were heading for the Burma Road. Agnew was sharp. He wouldn't like axing the pipeline extension, but would quickly accept why I was doing it.

Next, I looked at storage tanks and the handling of cargo delivered from Haifa to service stations and commercial companies. I made notes on recommended changes and left a copy with the installation manager.

From there I joined Dempster, the personnel expert, at branch headquarters. A large part of our job had to do with reorganizing lines of authority. Our objective was to decentralize authority, giving executives more mobility by setting up chain-of-command management responsibilities in sales, operations and accounting. A leaner team approach would make for faster responses to changing market demands and quick implementation. My job was to recommend reorganizations. Dempster's was to cut out dead-wood and replace it with promising junior staff.

Dempster's job was easy in Israel. There was a flood of competent immigrants, mostly Ashkenazi Jews from Europe with managerial experience in oil. Many were driving cabs, waiting for job openings on their level. From them I learned the Zionist view. The Jews, too, had a point. Why build a homeland in Europe where they had been persecuted for centuries? What better place for them, than the lands sacred to their shared heritage?

While Dempster interviewed the many qualified applicants for the few available jobs, I explored. I wanted to understand this upstart of a country that had galvanized the Arab world into a jihad.

Both population and industry was burgeoning. Tel Aviv was growing into a bustling metropolis and Haifa was becoming one of the most important ports on the Mediterranean. Heavy industry was springing up in both cities, and everything from chemicals to clothing was being manufactured. New Jerusalem, the capital and educational center of this new nation, was expanding into the hills, and everywhere housing was under construction to accommodate the deluge of immigrants that continued to pour in.

They came from nearby Yemen, Kurdistan, Iraq and Turkey. They came across northern Africa from the *mellahs* of Algeria, Morocco, Egypt and Tunisia. A warlike lost tribe of Jews in the Eastern Protectorate had fought its way to Aden. The wealthy Jewish community from South Africa came to Israel. Jews poured in from the refugee camps in Europe. They came from China and India, where they had settled 3,000 years before. They came from Australia, Canada and England, from the Argentine and the United States.

Some walked through burning deserts, others flew in rickety planes, packed beyond safe limits. Some came in deluxe liners. However they got here, these exiled and dispersed Jews, came from seventy-four nations to the one corner of the world where they would not be slandered or despised for being Jewish.

They lived in tent cities and tin shed villages that sprang up from the Galilee to the Negev Desert, and still they came. Israel's resources were strained to the breaking point by this influx of people from every possible background and circumstance, from the world's most sophisticated cities and from isolated desert communities where life had not changed since Old Testament days. Yet there was an air of optimism throughout Israel. They were welcomed here, and their new sense of belonging fired them with a dignity and drive that was making the impossible happen.

But the cities and villages told only part of the story and I wanted to know more. I arranged with Dov, a refinery worker from Tel Aviv, to visit one of the agricultural settlements which were springing up daily. Dov arranged the necessary permissions and, on his day off, I borrowed a jeep and we set out for a settlement just below the Gaza Strip, in the Negev Desert.

"Because my fiancée is there," Dov answered shyly, when I asked why he had chosen this particular settlement, so far from Tel Aviv. "She has just completed her basic army training and has volunteered to help construct this outpost. It is on the border, a combined farming and defensive settlement."

As we drove through the searing heat, Dov explained that the toughest of the frontiers was the Gaza Strip. Rebecca, his sweetheart, was helping to build a settlement within sight of armed Arab enemies.

"Ancient Gaza," Dov said, "where Sampson lifted the gates, now faces the gates of Palestine refugee camps. Gaza is the main base and training area for the *fellaheen*, mostly refugees whom the Egyptians keep caged like prisoners. They are fed very little food but much hate, then armed and turned onto Israel. These poor driven fanatics have done terrible things, including bombings of children's houses in the border settlements."

"Then why do your people stay?" I asked, mopping my sweat-soaked brow. Chilling as this news was, I wanted to keep the conversation going. It was somehow easier to think about these atrocities than to merely swelter helplessly in the 125-degree heat of the wilderness we were passing through. The endless slate fields and deep gorges and canyons of this wasteland were desolate enough, but mixed with the heat and the acrid taste of sand, they were making me nauseous.

Dov drove jauntily along, his tightly curled dark hair bouncing in the bright sunlight as we continued our bone-shattering ride. "Ask Rebecca," he grinned. "She will tell you why they stay."

We were on a narrow steep track that followed the course of a dry and winding creek bed, back into the hills. It was hardly more than a mule path and we bumped and jolted over ruts and lose stones until, rounding a shoulder of rock, we came upon the settlement.

While Dov and Rebecca greeted each other with shouts of joy, followed by hungry kisses, I looked around. Most of the settlement was still canvas. The water tank and irrigation pipes were under construction and a small patch of rocky ground was being plowed for gardens. Only a dining shack, a tool shed, and a pair of guard towers had been built. These few buildings stood in the center of a windswept bleak area that was defined by the rocks behind us and the outline of Gaza on the horizon. Emplacements of barbed wire and trenches faced the enemy.

We had driven all day. It was now dusk, and a day's work had ended. Dov and Rebecca, remembering my presence, guided me to the dining shack where the rest of the settlement, twenty young men and eighteen young women, were gathered. Our meal was spartan but spirits were high. The day's progress was reported and crews were arranged for the next day's work. Rebecca had to leave for guard duty, and she invited Dov and me to accompany her.

The desert moon rode low in a cold sky as we walked toward the forward outpost, right on the border marker. The day's heat was giving way to a windswept chill. The only sound was the wind, slithering through the sand.

We climbed over the embankment and dropped into the outpost trench that would be our only shelter through the bleak night. Rebecca peered out at the desert through the barbed wire, slowly and carefully noting each sand dune, each curve and shadow that the day's winds had formed. Satisfied, she leaned her rifle against the embankment and said to me, "And what brings you to our settlement, Mr. Hall?"

"He wants to know why you stay," Dov offered for me.

Rebecca frowned thoughtfully and brushed sand off her khakis. Removing her army cap, she shook loose a thick red braid of hair which fell down her muscular back. Finally she looked at me.

"And what else would I do?" she replied tartly.

I had the feeling that Dov had asked her this question himself, more than once. Two years older than Rebecca, this tall, sensitive young man had completed his military service and was working as a chemical technician at the Tel Aviv refinery. Dov, a builder more than a fighter, dreamed of a home and many children. He and Rebecca would have to delay their wedding until Rebecca's army service was completed, and Dov had confided to me how difficult this wait was for him. He also feared that her burning patriotism would lengthen their wait. Dov worried that she might choose to remain in this outpost past her army days, as some of its other members had.

"Well," I began, "you could complete your army service in some safer place. Dov tells me you are a nurse. Surely nurses are needed in the refugee camps near the cities."

"That is true," she replied, her head turning slowly as her eyes swept the surrounding landscape, "but there are many willing to work in those camps. Few volunteer for border settlements, and they are vital for Israel's survival."

"Aren't you afraid of the *fellaheen*?" I asked.

"Certainly, but my fear is mixed with pity and a need to show them truth. These miserable fanatics are being made pawns by the Arab leaders. On their own, they do not hate us. We hope to convince them by our calm and just presence that they have nothing to fear from us."

"Do you really think you can do that?" I asked, leaning against the embankment.

"The *fellaheen* have been formed by Arab leaders, out of the Palestine refugees whom the Arabs themselves created," Rebecca said quietly. "Had the Arab leaders obeyed the decision of the United Nations and adhered to the law, there would be no Arab refugee problem."

"After the 1947 partition, the Yishuv of Palestine begged the Palestine Arabs to remain. And after Israel declared her independence, she held out her hand in friendship to her Arab neighbors, even as they invaded our borders."

"Yes," Dov added. "Their avowal to murder the Jews and destroy Israel was the Arab answer to law and friendship."

Rebecca nodded, continuing to monitor the landscape. "Did you know, Mr. Hall, that most of the Palestine Arabs fled before we invaded? Racist leaders had filled them with fear, implanting in their minds for decades the fear of a massacre by the Jews. Those leaders played on their ignorance, superstition, and fanatic religious devotion. The leaders cared nothing for the *fellaheen*, only for their own personal ambitions. They betrayed their people. Blind fear and ignorance caused the flight of the Arabs. The Arabs who remained in Palestine were completely unharmed."

"Arab leaders wanted them out for political and military reasons," Dov said, blowing on his cold hands. "They planned to annihilate the Jews and didn't want a big Arab population around to clutter their operation."

"Exactly!" Rebecca agreed. "And they wanted to convince the world of Jewish cruelty by pointing to these refugees we supposedly forced from their homes. But only those Arab villages which fought against Israel were attacked, and there is documented proof that those who fought were promised they could return to their homes and loot a destroyed Israel after the Arabs had beaten us."

Silently digesting their conversation, I gazed out at the desert, which looked like a moonscape in the shadowy light.

"The Arab nations don't want these people," Rebecca accused. "They are kept caged like animals as a deliberate political weapon. In Gaza, the roads are mined and patrolled so that they cannot reach Egypt."

"And not a penny of the $200 million U.N. fund for their resettlement has been used," Dov angrily announced. "The Tigris-Euphrates Valley has some of the richest unused land in the world. It could take the entire half-million refugees, plus 10 million others!"

"Yes," Rebecca said bitterly, "while Israel, whose 7,000 square miles are half-desert, has taken in more than a half-million Jewish refugees from Arab countries and stands ready to take in more!"

Rebecca paced the length of the trench as she talked. "The fact is, the Palestine *fellaheen* were victimized by men who used them as tools, deserted them, then victimized them again. Kept penned up, fed with hatred, they are being used to keep Arab hatred of Israel at the boiling point. We hope to demonstrate, at these outposts, the truth of Israel: that we are law-abiding, peaceful people who wish them no harm but who are willing to defend to the death what is legally ours!"

We had talked far into the night, and I had been given a lot to think about. Feigning sleepiness, I crept back to the jeep, leaving Rebecca and Dov alone together for the few remaining hours until dawn.

I was awakened by the sound of the jeep's engine engaging. As Dov began the long drive back to Tel Aviv, I watched the day descend on the desert through a peach-colored sky and thought about the tribal and charismatic overtones of Middle East relationships.

♦ ♦ ♦

Dempster and I finished our work in Israel, two days ahead of schedule. Back in Cyprus, we decided to use these two days to check out the world's largest oil reserves. Since we could travel freely through the Arab states on our U.S. passports, we booked a flight for Saudi Arabia.

The few women who boarded the plane with us were black-veiled and shrouded in robes. I smiled to myself, remembering the high-fashion dark-eyed beauties who had boarded the plane with us in London. By the time we landed in Cyprus they had erased all traces of designer clothes and makeup, emerging on Cyprean soil in these same anonymous veils and robes. I had been told that Saudi Arabia still lived in the fourteenth century but could not imagine how, with the worldwide commercial ties that oil had brought, the country could escape modernization. I found out in Riyadh.

From the air, I saw a medieval walled city, surrounded by bright greenery and then stark desert. The airport was a mere sand strip with a windsock and a few old tents, where we were offered mint tea and coffee. We were on our way to Dhahran, but every plane required the king's express permission to land and take off. All passengers had to leave the plane and wait in the tents while a messenger drove to town to find out if King Abdul Aziz required the plane to transport any of his family or possessions.

A curious prolonged sound filled the air. I looked outside and discovered it was the creaking of wooden water-wheels, sighing and gurgling as blindfolded camels plodded endlessly in circles, dragging leather buckets from the earthen wells. This sound had been part of this desert settlement for centuries.

I listened to it for hours, waiting for the messenger's return. Twice during our wait, I heard the cry, "*Allahu Akbar*" (God is most great). Immediately following the call, all the Muslim men in the tent rose, kicked off their leather-thonged sandals and, facing Mecca, began murmuring their prayers.

"The entire country," a Muslim fellow traveller told me, "stops everything for twenty minutes five times a day to kneel down and worship. The Kingdom was created in the name of God."

"In the name of God" appears on every official paper in the Kingdom. Pilots welcome you on board with it. In the name of God, thieves have their hands chopped off and murderers and rapists are executed in the streets. Proven adulterers, regardless of rank, are similarly executed.

I am beginning to get a picture of Muslim piety. It excludes the drinking of alcohol and includes a fateful acceptance of "God's will." Mecca and Medina, where God revealed the true religion to Muhammad 1,400 years ago, are more important to Saudi Arabia than is its oil. One thing the Kingdom stands for above all else is Islam, which pervades every moment of every day in the lives of faithful Muslims. There is much bad feeling here about Jerusalem, the third holy place for Muslims, being occupied by Jews.

"Our hatred of the Jews," the Muslim traveller tells me, "dates from God's condemnation of them for their persecution and rejection of *Isa* (Jesus Christ) and their later rejection of His prophet, Muhammad."

"But," I ask, "is this reason enough to kill them?"

"To kill or be killed by a Jew in war ensures Muslims immediate entry into heaven," he answers.

He finds it difficult to understand how Christian powers like Britain and the United States can want to help and reward these Jews who mistreated *Isa*. He sees Arabian support of Palestine Arabs as natural and pious, and is puzzled by our failure to understand and accept this expression of their faith.

After Riyadh, Dhahran seems like an optical illusion. There in the middle of the desert, on the edge of the oil fields, it looks like a small Texas town. Neat bungalows with creepers up the walls and green lawns are laid out in rows, like an instant suburb. There is a baseball park, a movie theater, swimming pools and tennis courts. But the suburban illusion is broken at the edge of town. A high barbed-wire fence surrounds this isolated outpost, enclosing the "Aramcons," as they call themselves.

A lot of the residents are, in fact, from Texas. But they have acquired a special character all their own. As one of the engineers told me, "Aramco is the nervous child of four parents, subject to the whims and jealousies of each. There's a sense of uneasiness here, increased by the Arabs' anger at American policy toward Israel and the eccentric demands of the King.

"We're just providing technology and oil for the good of the world. We're proud of that, and proud of our contribution to Saudi's social services and industry. God knows we want to avoid the paternal and imperialistic attitudes of BP in Iran and Kuwait! But no matter how we try to avoid politics, we are a Western enclave in the middle of an autocratic kingdom. So here we are, champions of the Arabs, from a country committed to supporting Israel, working in a very special relationship between a company and a country. That's why we have the fence."

What he said made sense. Their only safety was in mistrust because, as another Aramcon had explained to me, the desert code still applied. A guest was sacred while he ate your bread, but once he was out of your camp he was again legitimate prey.

I was glad to get back on a plane and out of Saudi. We spent only one day in Lebanon, evaluating facilities in Beirut and Aman, before heading for the Area's head office in Cairo. Egypt was a surprise.

Chapter XIX

The Dark Continent

Like the rest of Africa, Egypt is primarily Muslim. Cairo, with its population of two and a half million in 1951, was the largest city in the Muslim world, but seemed more European than Arab or African.

Modern air-conditioned apartment buildings lined the Nile like cliffs. In the crowded alley ways, neon lights highlighted the cafes and bars, packed with Egyptians discussing King Farouk's recent denunciation of the 1936 treaty with Britain. Britain had protested, but was helpless to alter Farouk's assumption of title as King of the Sudan as well as of Egypt. Nobody was going to mess around with the man who controlled the Suez Canal, through which oil travels out of Saudi Arabia.

Around the world, people were nervously monitoring the increasingly close relationship between King Farouk and King Abdul Aziz.

In Cairo cafes, King Farouk, despite his bold move to annex the Sudan, was getting less praise than Colonel Gamal Abdul Nasser, who was declaring the Arabs to be one nation. Nasser's blend of nationalism and radicalism seemed to intoxicate the Egyptians. Still burning from their 1948 humiliation by Israel, they lit up with pride and ambition when they spoke of Nasser.

Cairo became our home base. We spent the next six weeks analyzing operations in Suez, Port Said, Ismailia, and lesser installations. Small passenger planes were made available to us for short hops to desert towns, of which there are many.

Egypt is twice the size of California, but ninety-six percent of it is uninhabitable desert. My job dealt with products and marketing and most towns we visited were in the Nile valley, along the thick, fruitful silt of its Delta, or near the few scattered oases controlled by rich pashas. My job dealt with products and marketing. It was the engineers and geologists, searching for new oil deposits, who roamed the desert.

Most of the people I saw in Egypt were oil personnel, and soon, I stopped sorting them by company. The position of Shell Company of Egypt in the Group was similar to that of other Group marketing companies in the Consolidated Area. Its shares were held half by Shell Petroleum and half by Anglo-Iranian Oil. Shell was predominant among the four major marketing companies and one of three oil companies then marketing in Egypt. All four marketing companies supplied the Government Refinery.

So many refining and shipping facilities were jointly owned and operated that it was futile to think in terms of "ours" and "theirs." The oil business in Africa is a strange combination of cooperative marketing of product and competitive exploration to discover and get the rights to new oil sources. Whatever efficiencies Dempster and I could bring about would benefit all the oil companies holding drilling rights in Africa.

Between various site visits, Dempster and I would attend the lunches and evening parties for overseas visitors given by the oil companies operating in Egypt. I often ran into people I had met in London or New York.

In the forty days of our Cairo stay, we attended about thirty dinners and parties. Dressing for dinner, a mandatory relic from the days of British rule, was rarely uncomfortable, as the spring weather was usually pleasant. During the occasional hot spells, hostesses were merciful and seldom made us suffer past the second drink. The call of "Gentlemen, you may remove your jackets," brought sighs of relief. We would quickly shed our black mohair jackets revealing our cummerbunds and white cotton shirts.

But the heat of Egypt seemed arctic compared to that of the Sudan, our next stop. At the confluence of the Blue Nile and the White Nile, Khartoum is a nexus of trade and contact between Mediterranean, European, and Arab-influenced Africa and the rest of the continent. And it is one of the hottest places on the planet.

In Khartoum I could feel the heat from the street burning through the soles of my shoes. I was glad we would not be here in June, when the *haboob* winds were said to blow desert dust "as black as smoke from an oil fire" over the parched towns. Now, in March, the sight of thousands of cranes and storks along the Nile made the heat bearable.

The Nile keeps the entire Sudan alive. This river, the longest, loneliest, most powerful river in the world, is actually made up of several rivers each named the Nile. The main branch, with its source in Uganda at Lake Victoria and fed by mountain rivers originating in Tanganyika and the Congo, is known as the Victoria Nile for a short space and then becomes the White Nile. Looping north into the Sudan, it spreads out as if almost lost in the immense Sudd swamps, then straightens out and flows as a giant conduit through the whole of Sudan.

The Blue Nile rises in Ethiopia and meets the White Nile at Khartoum. From there, as the main Nile, it pushes northward into Egypt and it flows powerfully for 1,900 miles to the Delta. Glad as I was to be missing the hottest months, I regretted missing the sight of the "miracle" of the Nile in August and September, when it swells up like a cobra and spills into the burning Egyptian desert, turning thousands of square miles of land into lakes which dry up a few months later every year.

Of the total Nile flow, only a fifth is immediately usable. Another three-fifths can be stored, but the remaining fifth is lost in its unharnessed charge down to the Mediterranean, carrying with it millions of tons of real estate.

To further harness the Nile is one of Africa's greatest challenges. As an engineer I can appreciate the difficulties. In the whole of the Sudan, for instance, where the Nile runs to 2,144 miles, only four bridges cross the Nile, clustered in one small area around Khartoum. There is no bridge at all between Owen Falls in Uganda and Kosti, 198 miles south of Khartoum, a distance of almost 3,000 miles.

King Farouk's annexation of the Sudan was motivated by his perceived need to safeguard the Nile; Egypt's only source of water.

Though they are neighbors, the Sudan, four times the size of Texas, is quite different from Egypt. Sudan's people are of Arab and Negro ancestry, superimposed on an older Hamitic stock. About a million are urban and Western educated, and they seem to despise the "savage" nomads and tribesmen of the desert and equatorial jungle. These "savages" are physically magnificent and have their own high standards of conduct and honor. They seem to enjoy a happy way of life.

These various groups of desert nomads and jungle tribesmen seemed to have little political consciousness, but they definitely possessed a great sense of superiority and racial pride. They tended to patronize British officials for the "unfortunate whiteness of their skins."

One day I saw a male Dinka near Khartoum. Stark naked, he stood on one leg like a stork supported by a long stave. Beads of artful scar tissue adorned his body. I wanted to speak with him about the Dinka ability with cattle, but I did not know his language. There are 264 Sudanese languages, many of which have spread throughout Africa. The Masai of Kenya and the giant Watusi in Ruanda-Urundi are cousins to the Hamitic Sudanese.

The 6,500,000 northern Sudanese are mostly Arabic-speaking Muslims, strongly under Egyptian influence. They belong to Europe and the Levant. The 2,500,000 southern Sudanese are pagan, primitive, darker-skinned, and speak their own languages. In all the Sudan, Khartoum, Wadi, Port Sudan, and Juban were the only places a visitor accustomed to European amenities could stay with any comfort. It was not until the year of my visit, 1951, that the U.S. had sent diplomatic representatives to Khartoum.

It seemed certain that some day this mixed bag of Sudanese people would demand self-government. I sensed an intense dynamism among the people. One young Sudanese, working at the Port Sudan facility, confided to me that his country wants to combine the best of Africa and Europe. Good roads, schools and hospitals were a must, but they also wanted good movies. With its current favorable balance of trade from cotton and gum arabic, all of this seemed possible for the Sudan.

The country was growing in sophistication. All nine provincial governors were Sudanese and the civil service was being run by 6,000 Sudanese to only 1,000 British.

Talking with workers in Khartoum, I learned that the Sudanese liked the Egyptians, but did not respect them. Conversely, they respected but disliked the British. After a century of subjugation by the British and Egyptians, the Sudanese seemed to be preparing for freedom.

Chapter XX

Moving South

Our journey through East Africa took us through Kenya, Uganda, and Rhodesia. It was the end of the rainy season, cool, often overcast in the highlands but sticky and tropical along the coast and in Uganda. During our travels, we saw lions, leopards, elephants, giraffes, and other wild animals, plus the spectacular Victoria Falls.

Political tensions appeared to ease as we moved south. Britain still controlled East Africa, and the oil facilities we inspected were all in good, fairly modern condition. Most were English or American built, needing only minimal technological update. Many of the refining machines in Africa were behind the times, unable to handle aviation fuels and behind in auto refining abilities.

But the pipelines and shipping facilities needed urgent updating, similar to what was done to England's during World War II. The difference was that England, unlike Africa, had begun planning its improvements some ten years before the war, and England's political stability had made long-range planning easier. Africa's difficult topography, including mountain ranges and deserts, meant that it was going to be costly to bring tankers, railroads, and pipelines up to competitive standards here.

Dempster and I spent a lot of time with local managers, discussing ways around budget, supply, and manpower limitations. In some cases we recommended trips to the States for them as quick courses in modern technology.

Dempster needed an extra week in Kenya to handle some complex shuffling of personnel, so I took some vacation days and went fishing. The travel pace, the intensity of the facilities evaluations, and the political considerations that kept surfacing at the endless cocktail parties were wearing me down.

The fishing resort in the Kenyan mountains was a welcome relief from these pressures. I was restored by the combination of tropical climate and bracing altitude. The frontier spirit of the place was intensified by the dramatic beauty of the landscape and the brilliant blue sky. It was a great satisfaction to catch a few fish in the pools that were dammed up along the resort's creeks, then enjoy a beer on the veranda while these fish were being prepared for my dinner. I met a couple of European settlers and an aide to the British Governor of Kenya, and learned that government here, too, was in flux.

Alfie, a rugged, hardworking pioneer who employed many African laborers on his tea and coffee acreages, did not hate Africans, he said.

"I actually feel a lot of affection for the blighters," he told me. "But they have to be kept in line, like children who need an occasional spanking."

"Rubbish!" Nigel broke in. "They are vicious primitives who will destroy everything we've built here if we don't rule with an iron fist."

I looked at Bruce, the governor's aide, who was staring off into the distance. For several moments he said nothing. Then, turning to us, he remarked calmly, "I think you both underestimate the tempo of African development. These people are organizing on a substantial scale. The Kikuyu will soon represent a nationalistic movement and the only way we're going to maintain a white presence is to help them toward self-government. Because of thinking like yours, they don't even have a vote yet, much less opportunities for education or government jobs. It's going to catch up with you, sooner than you care to think."

"Ridiculous," Nigel snorted. "Give up rule, they'd only return to their *blackness*. As it is, we consult the damned Africans too much."

"Oh, I think in time they could grow into self-government," Alfie said. "I have camaraderie now with my field workers. They feel quite free to talk up to me, make suggestions and such. A few of the brighter ones have even moved up to being crew bosses. Yes, I think there's hope, but it will take a long time."

"We don't have a long time!" Bruce insisted. "You know as well as I that they're not ready, but they think they are. And we'd better give them opportunities now. Once they get a taste of governing they'll begin to see how complex it is, be more willing to take the time to learn from us. If we don't allow them to begin, they'll take over without our help. And you settlers will be in real trouble."

"Oh come, come," Alfie chuckled. "They look to us for guidance and protection. They're a carefree lot, don't want the burden of self-rule. If they did, they could have run us out long ago. Just look at the military—all blacks, with not one whit of resentment toward their white officers. They consider it an honor to serve the Crown. Why, the black boys here are as eager to get into the King's African Rifles as American boys are to get into West Point."

"Terrible mistake, arming them. Not stable. Could turn on us any moment," Nigel growled. "And you Americans aren't much help," he added, glaring at me. "You've lost our empire for us because of your dollar pressures."

Nigel's accusation certainly applied in Saudi, but the British in Africa looked secure. I thought about the American oil company doing a petroleum survey in northern Kenya and the auto assembly plant planned in Mombasa. Other than these enterprises, Coca Cola, and the safari business, I didn't see any U.S. presence in Kenya.

"It's not the Americans," Bruce argued. "It's the economy. The poor British taxpayer is as much exploited for Kenya as the natives are. We're working on a Ten-Year Development Plan but we're dependent on Britain to fund it. I hope the Americans strike oil, create jobs, and pay taxes here. At the rate money trickles in now, we'll never beat the clock."

Our conversation was interrupted by a call to dinner. Over the next few days we had ample opportunity to pick up where we had left off, and I learned a lot about what went on below the orderly surface of this country.

Three sides were competing in the struggle for power in Africa, the British, the Africans, and people of Indian descent. The Brits were divided between the die-hards like Nigel, wanting permanent white supremacy at any cost, and the government and moderate white settlers, who favored a pluralistic society.

Kenya is mostly arid and is chronically poor. As a British colony, it was being regularly bailed out by London. Its three geographically distinct areas consisted of the steamy coastal strip, which Britain leased from the Sultan of Zanzibar, the Northern Frontier Province, a desert wasteland, and the central plateau, with altitudes ranging from 3,000 to 10,000 feet. The plateau's rich, fertile "White Highlands" covered 16,000 square miles and were the stronghold of European settlement.

Kenya lacks rain and river water but is in the midst of the great African lakes that include Victoria, which Kenya shares with Tanganyika and Uganda. Associated with the great African lakes is the Rift Valley, which stretches 4,000 miles from Israel's Jordan Valley across the Red Sea into Ethiopia and south through Kenya. The Rift Valley lies naked and exposed near Nairobi. In this giant valley, volcanoes lie like pebbles in a ditch, but the ditch is as wide as the distance between Washington and Baltimore.

Kenya lies on the Indian Ocean and has been involved in seafaring trade for at least 3,000 years. Arabs, Chinese and Portuguese have traded along its coast, but only the Arab slave traders penetrated inland. In 1895, twenty-six years after the Suez Canal opened, Kenya became part of the British Empire.

In 1903, before colonization by European settlers was fully under way, the British government offered the Kenyan Highlands as a National Home for the Jews. Zionist leaders in London rejected the offer, despite the fertile land where two harvests a year are possible, the sunshine brilliant and the climate bracing, with malaria and sleeping sickness unknown and the tsetse fly rare. That decision may have changed the course of history for both the Middle East and Africa.

At the time of my visit, a handful of Europeans owned a quarter of this land in Kenya, and five and a half-million Africans got along on the remainder. The African population had doubled in the past twenty-five years, creating overcrowding and unrest among the Masai, Kikuyu and Kamba tribes who were restricted to "reserves" set aside for them in and around the fertile White Highlands.

Nairobi, a chic little city with an international smart set, was the safari capital of the world. Only six miles from the nearest Kikuyu reserve, Nairobi seemed oblivious to the growing Mau Mau rebellion.

Nairobi is also an Indian town. Indians controlled retail trade and made up a large part of the professions and skilled trades. This group has been caught in the middle in Kenya, afforded higher status than the Africans but lower than the British.

The British feared that, with their increasing closeness, the Indians would provide the under-educated Africans with an informed and capable leadership. Believing the Africans incapable of organizing, the Brits assumed the Indians were behind the Mau Mau movement. Less concerned about revolt, the British were worried that the Indians would teach Africans the techniques of passive resistance and civil disobedience which Ghandi had used in India, proving that an unarmed colonial people could win a revolutionary struggle.

The color bar was a paramount issue in Kenya, where there was three-way segregation. In the Nairobi airport there were separate toilets for "European Gentlemen" and "Asian Gentlemen" and none at all for Africans. The enormous African majority on the bottom rung was asked to be content as chauffeurs, gardeners or houseboys, with no share in the development of their country. But the Africans wanted the same thing as the white settlers—more self-government.

Britain, presently retired with dignity from Iraq, Jordan and Israel, had helped the Union of South Africa develop into a self-governing Commonwealth, and made great concessions in Egypt and the Sudan. Britain was committed to Dominion Rule in Kenya, but only when they were sure the Kenyans could resist tribalism and Communism.

Yet, it was going to be impossible for the British to continue to rule without their full consent. Overwhelmingly outnumbered, the costs of trying would be high. Yet strangely enough, London, where a number of organizations were devoted to anti-imperialistic causes, remained the capital of Africa, politically as well as economically.

I continued to think about this situation as I fished. An image of a tepee came to me, with the white settlers, the Africans and the Indians each forming one of the interdependent poles that would allow the structure to stand, covered over by the hide of British rule that, so far, had kept out the winds of chaos. But the reality was that change was in the air. It looked to me like the changes were going to be powerful enough to tear holes in the fabric of Kenyan society.

Chapter XXI

Color Bar

In early June, we flew to Johannesburg, South Africa, where it was mid-winter. The days were fine and bright, the nights cold in the high veld. We enjoyed California-like weather in Natal province, while rain and cold weather greeted us at the Cape, where Shell's head office is located.

The head office is responsible for the activities of the four distribution companies operating in the Union and also acts as agent for Anglo-Iranian oil. We visited all twelve branch offices and recommended extensive reorganization in each, in order to achieve more aggressive marketing. Dealer education programs were outlined, and a pipeline was suggested between Durban and Wadeville, along the Durban-Johannesburg highway.

Additional pipelines seemed advisable. Long-distance hauling by road was non-existent in the Union, and most haulage was done through railroads, harbors, or airways, with all three controlled by government.

The government kept coming up in all discussions about reorganization. A ripe situation existed for Shell to assume marketing leadership, provided Shell could oust Vacuum from its most favored position with the government. Local Shell executives who knew the political ropes suggested we build public acceptance through progressive steps. They also suggested contributing staff assistance to public projects and government commissions as a way to gain the most favored status with the authorities.

It soon became clear that Shell's market advances would depend on currying favor with the government. To determine strategy, I had to learn more about how the Union governed itself. This turned out to be a complicated and discouraging chore.

The Union of South Africa was a strange nation, governed exclusively by whites of European ancestry. It was the only independent country in Africa run by whites, with more than half the entire white population of the continent. The Union was fully sovereign, independent and a self-governing member of the British Commonwealth, with its legislative capital in Cape Town, judicial capital in Bloemfontein, and its administrative capital in Pretoria. Although the tie between London and Pretoria was mostly formal, there was fierce sentiment among many Afrikaners to secede from the Commonwealth and form a republic.

Certainly the politics were different. One of the main characteristics of British rule in Africa was its policy of training their Africans for eventual self-government within the Commonwealth, but there was none of that in South Africa. White supremacy was the rule, and blacks, comprising four-fifths of the population, had no say.

Statisticians were saying that the European population might double in the next thirty years, but that the African population might triple. The numbers were on the side of the non-whites in the long run, but for the time being, nearly ten million black and brown people were denied the most elementary rights and privileges by a white minority that was divided among itself.

The Union, formed in 1910, consisted of four provinces. The Cape Province and Natal were former British colonies and British influence was still strong. The Transvaal and the Orange Free State, formerly independent Boer republics, were always the stronghold of the Afrikaners, descendants of Dutch and other non-British Europeans. The Union was officially bilingual, with both English and the Dutch-based Afrikaans as its official languages. The British and Afrikaners were firmly divided by their backgrounds, speech, and intense economic and political differences.

The British were the merchants and traders. They controlled industry, the mines, and the banks. The Afrikaners stayed on the land, or went into the professions or government service.

For the past forty years, political power here had been concentrated in the hands of a small number of men, most of them Afrikaners. The Broederbond, a secret Afrikaner organization, exercised most of the political power behind the scenes.

The Broederbond stood for an Afrikaner republic, with all links to Britain abolished and giving any English subjects who choose to stay "inferior citizen" status. The Broederbond was noted for hatred of Freemasons, Catholics, and Jews, and for its blistering anti-black sentiments. Though membership was secret, it was common knowledge that two-thirds of the candidates who stood for parliament in 1948 were members.

As a self-governing member of the British Commonwealth, the Union had a Governor General, Dr. Jan Smuts, who was both a representative of the British Crown and an avowed republican. Smuts' Nationalist government had recently passed acts requiring the entire population to be classified into racial groups and prohibiting marriages between Europeans and Africans and sexual relations between Europeans and Indians, mixed-race people or Africans. Locally, Afrikaners referred to African people as "Natives" or, more derogatorily, as "Kaffirs." Indians were called "coolies."

South Africa was run by two groups who hated each other. One was political, the National Party, consisting of Afrikaner nationalists, including the Broederbond, the Dutch Reformed Church, the universities, and the Governor General. The other powerful group was the Chamber of Mines, composed of financial interests mostly concentrated in London. British capital had no choice but to accept Afrikaner political domination, and there were plenty of British who, wanting cheap black labor for industry and the mines and not necessarily objecting to the practices of racial discrimination, favored some form of apartheid.

South Africa was the richest and most industrially advanced nation on the continent. With its gold, diamonds, and uranium, plus two-thirds of all the railroad mileage in Africa and a thriving industrial base, it was a great economic power and a very important customer for British exports. It was also Britain's chief source of gold, and the United Kingdom had invested heavily there. England considered the sea route around the Cape of Good Hope strategically important. It served British interests to be on good terms with South Africa, despite the avowed British dislike of apartheid and fear that the Union would leave the British Commonwealth.

Most of the oil men here were Union citizens who seemed to endure the social and political tensions through an endless round of social gatherings. In Pretoria, the administrative capital, and therefore the capital of Afrikaner nationalism, the cocktail parties were tense. Conversations always returned to politics. It was common to hear defenses of apartheid.

"Once we give them political rights, we'll be faced with eventual rule by the majority. So we can't give them political rights."

"You have apartheid in the United States based on money. In England it's based on class. You have no right to criticize our form of it. We think it's the best policy."

In Johannesburg, predominantly British and the most important city south of the Sahara, the tone was either more pragmatic or genuinely worried about the color bar, for a variety of reasons.

From a bearded professor, I heard that "this is the only country in Africa with active Communism, and it's no surprise. Communism can appeal to anyone who has a grievance against society, and there's not a single educated African in this country who doesn't feel hindered and frustrated."

"True enough," agreed an accountant, "but who's going to support them now that the government has passed the Suppression of Communism Act? The minister of justice has only to declare a man a Communist in order to fine him, ban him from public life, or put him in jail for five years. And all it takes to be 'declared' is to want any kind of political, industrial, social or economic change."

A branch manager had wandered over, balancing a cocktail in one hand and his pipe in the other. His wife followed close behind him.

"Quite!" he exclaimed. "So anybody who says that housing conditions are bad in the locations thereby influences Africans to protest against them and can be declared a Communist. Best to keep one's mouth shut and one's gun loaded."

"Oh, God, I'm so sick of the crime and violence," his wife moaned, turning to the younger couple that had joined us.

"Why shouldn't the crime rate be high?" the younger man asked. "Why shouldn't they steal from us? Look how we make them live, in those damned 'locations!' Fenced shanty towns with no electricity, water, drainage or health facilities . . . it's no wonder more than half the children die before age two."

"And what kind of life do they have if they live?" his journalist wife asked. "Houses of cardboard, wood and tin they aren't even allowed to own . . . no trees, no phones, cement floors that give them rheumatism. We require them to be inside from dusk to dawn, and we provide the poorest transportation. Why, many of them spend four or five hours a day getting to and from their jobs in town."

"But in locations at least workers are with their families," the first woman pointed out plaintively. "Those that live and work in white households must be single. Mine workers have to leave their families behind and live in migratory barracks for the entire two years of their contracts."

"But what's worse," the liberal young husband broke in, "is their political position. It's illegal for mine workers to quit or leave their jobs or to strike."

"That's no worse than the position of the Africans in towns," his wife insisted. "They have to carry four different passes at all times. They have no mobility without permission, but Natives—even whole tribes—may be moved from one area to another by administrative order. No court authority necessary. And they may be arrested and held in jail for three months without charges."

"Yes," her husband added. "During which time they get 'rented' out as cheap labor, and never see a penny of it. I wouldn't blame them if they went Communist, or went back to their witch doctors or turned into Mau Maus. The situation is abominable."

I had to agree. These startling revelations left me feeling I had been doused with green slime. It seemed clear that the Africans here, as in Kenya, were nowhere near ready to control or run a modern state. But how long would they accept their inhumane subjugation by a white minority?

I noticed Dempster unobtrusively assessing some of these people who were oil personnel whom he would be reshuffling. Dempster found these parties useful but I hated them. I just wanted to get the job done and get back to England. I said good night. In my hotel room a little later, I placed a call to Virginia in Rome.

Chapter XXII

The Road Home

My evaluations and reports finally completed, I met Virginia for a week in Rome. I needed the civilizing influence of this ancient city, and Virginia's exuberance, to get the taste of South Africa out of my mouth before seeing Mayne again.

Through the entire trip I had been thinking about Bousfield's suggestion that I get back into the engineering end of the executive path. But what I had seen in the Middle East and Africa convinced me I didn't have the stomach for it.

Politics weren't my forte, nor were the complicated intrigues in foreign countries. I was tired of separations from my family, and unwilling to move them to any of the hotbeds of colonial politics I had just visited.

My engineering skills had been an important part of this trip. I had enjoyed making all those operational recommendations, but found myself much more enthusiastic about Sales and Marketing solutions in the Consolidated Areas. For better or worse, I decided, I was a marketing man. As for the board of directors, I'd just have to let the chips fall where they may.

We had a leisurely week in Rome. We visited the Vatican, dined with old Shell friends from the States and Europe, and did the usual tourist things.

Virginia was very subdued. She had brought a cablegram from John to Rome which had us both upset. John had dropped out of his engineering program at Purdue to join the Marines. The Korean conflict was on. We had known he might be drafted, but his going in as an enlisted man worried us. It only increased his chances of being killed. We comforted each other as best we could, and, in our separate ways, worked at accepting the terrible possibilities that were beyond our control.

I think I was better prepared to make this acceptance than Virginia was. For her, John remained a precious child. The click of her maternal time clock had not kicked in, releasing him from her habit of protective care. It probably would not happen until he married. John sometimes joked with her about her fussing over him, though for the most part he accepted it with good grace. With me he was much more explicit.

Late last summer, a few weeks before his return to Purdue for his sophomore year, I had been working on the gazebo in the backyard while John was putting up the storm windows, as he had done for the past several years. I had just gotten home from a dealer recruitment trip and was still in high gear that Saturday morning, not yet shed of my executive mode. Without thinking, I began overseeing John's storm window installations. John quietly took it for about ten minutes. Then he put down his tools and walked over to me. He planted his feet about two feet in front of mine. John, too, was now six feet tall. We stood eyeball-to-eyeball, and I saw his anger.

"You must be a real son-of-a-bitch to work for if you have to tell me how to put up storm windows," he said. "But I don't work for you, and I don't have to take it." Having said that, he turned and went back to his work.

His back to me, I watched him work for a few minutes. He moved surely and confidently, with the efficiency of long practice. Then I just watched him. When had his shoulders broadened? When had his muscles developed? This was no longer the scrawny, stumblebum kid I was used to.

And so I realized that John was growing up. I also realized that, in my concentration on his project, I had neglected his feelings that day. I didn't say anything to him but promised myself I would begin treating him with respect and dignity. He had let me know he considered himself a grown man and he was entitled to be treated as such.

This change of attitude didn't come easily for me. Habits of long duration are hard to break, especially in close relationships. I was still working on it when news of his enlistment reached me. But I'd had one more year than Virginia to get used to the idea that John was old enough to run his own life, and I tried to share this new view with her. It was a comfort to both of us that Katie was safe in Montclair, with old and trusted friends.

Virginia and I arrived in London on August 4, where I teamed up again with Dempster for three weeks of conferences with Group executives about our findings. My main point was that control in the Consolidated Areas was too centralized.

"Give local managers financial authority through authorized expenditures," I recommended, "instead of holding them up with bureaucratic paperwork that often costs more than the item they're waiting for. It will save London a lot of paperwork, too. Let them handle sixty to seventy-five percent of the transactions. If they can't do that, they don't belong on the job."

"Have senior, well-trained people in the field," Dempster added, explaining how Aramco did it in Saudi, where everybody who amounted to anything had been given overseas training. All Saudi managers had bachelor's degrees, and all had been trained in either the States or in Europe.

We submitted our summary reports containing the broad outlines of the plans we'd hammered out with the local managers. These were matched with detailed plans that had already reached London from these managers, based on our collaborations in the field. London studied all plans, challenged our ideas with probing questions, and finally agreed to implement them.

Now that we had a green light, Dempster and I flew back to Cairo for a week. This gave George time to facilitate his personnel changes and gave me a chance to check the pipelines and storage facilities. The Suez Canal was the portal through which increasing amounts of Middle East oil would flow. With the pressures to further open up immigration to Israel, we expected a big jump in consumer business. This area would be expanding rapidly under the new plan, calling for a lot of new personnel. I had to see if the facilities there could be further modernized to handle the growth.

Dempster and his wife flew home in August, but at Mayne's request, Virginia and I remained in London for another month. It was a working month, but generously interspersed with cocktail parties, sight-seeing, golfing, and theater. Even the business dealings seemed leisurely, consisting of trips to British pipeline stations and long luncheons with executives of Shell-Mex and British Petroleum.

Mayne was impressed with my Middle East and African work, and was grooming me to be a contact person between U.S. Shell and executives of the international Group. I was a logical choice, since I was better acquainted with more overseas Shell people than anyone else in New York, and had gained firsthand experience with some of the international complexities.

This grooming didn't end in London. In early September, Virginia and I boarded the Queen Elizabeth for a month in Europe. We went to France, Belgium, the Hague, Denmark, Switzerland and Finland, visiting branches which had requested my services to do for them what I'd done for the Consolidated Areas. It turned out to be an informal reunion as I was greeted in many branches by people who had attended my classes in the States.

Mayne's purpose in sending me was to let me see how the Europeans worked. They would be sent to the States to learn how we did things. I would be one of the people in the States they would be working with, and knowledge of their operations would help me understand how American expertise could be translated into European efficiency. I would be their American guide and host, as they were now guiding and hosting me in their countries.

Wherever we went in Europe, people expressed concern that East Germany might be arming for another attack. Much of my work involved advising how oil could work to supply these countries' armed forces.

In France, I met an old friend from World War II, Louis Devoux, President of Shell Français. He was a former army officer in charge of the French underground, and had been in London while the Germans occupied France. He had helped us coordinate the pipeline and the attack on Normandy with the underground. He firmly believed that the Germans weren't through with France, and wanted Shell Français to be as prepared as Shell had been in Britain for World War II.

Our present work seemed like old times, generously ladled with sumptuous French hospitality. Devoux and Virginia took to each other immediately, often getting into cozy chats about our children. His view of John's enlistment as an act of heroism did much to comfort her and then he heard of Katie's plans to major in French in college next year, he said:

"*Trés bien*. It is the language of culture. And did you know, *mon ami*, that I do my little part to spread the appreciation of this great language? *Oui*, each year I sponsor twenty daughters of Shell people to spend a year in France, living with French families to get to know French as it is spoken and the culture as it is lived. Would your *juene fille* like to be one of these?"

With the old sparkle returning to her eyes, Virginia replied, "*Mais oui!*"

On October 5, we were given a farewell party in London's St. Helen's Court and a week later we were hugging Katie at the Idylwild airport and telling her about her forthcoming trip to France.

Chapter XXIII

Calm Before the Storm

For several years, back in New York, I continued to see many of the people I had met on my 1951 trip. There was a constant stream of overseas personnel visiting New York, attending classes taught by me and other specialists.

Mayne and Bousfield had correctly foreseen the domino effect of Venezuela's fifty-fifty demand. Saudi had soon followed and, though Mossadegh's attempt to nationalize British Petroleum's holdings in Iran had failed, there were rumblings of nationalization from other oil-producing countries.

Nasser overthrew the Egyptian monarchy in 1952. He was urging the Arab states to unite in cleansing their world of Western exploitation. He declared that the "Arab nation" had been subdued by the Ottoman Turks, then divided against each other by the Western imperialist powers. The American-supported creation of Israel was the latest example of this. It was time, Nasser said, for the Arabs to unite. They seemed to agree with him, though for diverse reasons. The oil companies were getting nervous.

The Egyptians wanted Israel destroyed, but they lacked the money and resources to do it. The Syrian socialists wanted to get rid of the little king of Jordan who was a friend of the British and a symbol of outdated tribal monarchy. The Jordanians wanted a highway to the sea and a port on the Mediterranean. The Lebanese wanted money and trade, and the Russians wanted a socialist arc from Baghdad to the Athens. For each of them, the Palestinian Liberation Organization (PLO) had a peculiar use. They could praise it publicly or damn it in secret, and pay generously to keep it alive. Either way, they all needed money to fund their individual wants, and all knew that oil was their best source for more money. And Venezuela was making sure that each Middle Eastern country knew that nationalizing oil fields was the surest way to an immediate jump in income.

With more and more oil fields likely to be nationalized, the Group had set its sights on the control of transportation and sales rather than the ownership of crude. This meant that Group operations throughout the world were being tightened up and given more competitive muscle, and U.S. Shell was the model.

It turned out I had been chosen for the Consolidated Areas job for certain reasons that Bousfield had not mentioned. Back in New York, with my new international understanding of the business, my boss informed me that the trip had been part of a process to groom me for heading up the U.S. Marketing Division. Bousfield had been correct on one point: an operations man with a technical background was needed for the job. My new assignment was to consolidate the day-to-day operations for all of U.S. Marketing, which was now plagued by a new source of competition, the independent oil companies.

We were tightening up stateside operations, too, to combat this new problem. It was a situation of our own making, one of those solutions that later become a problem.

During World War II, all the major oil companies had increased their refining capacities. After the war we needed new markets for our increased production, and had begun selling to independents. Until then, they had competed with us only on price, selling lower quality gas from bare-bones stations often in inconvenient locations. Now they had gas as good as ours (It *was* ours!) and began upgrading their operations. They moved into better sites and tempted customers with improved stations, giveaways and premiums, while maintaining their lower prices. They grew and grew.

We fought back by meeting their prices, but to do that we had to achieve economies in other areas. It was time to stop basking in the growing market and tax advantages and to fight seriously for our market share. As long as the "independents'" marketing margins were favorable, I reasoned, they would continue to grow. We needed a careful plan for marketing: real estate deals that would let the dealers prosper, improved efficiency and effectiveness at every level of distribution and help for jobbers so they could accomplish the same thing in their franchised areas. We should begin, I decided, by dropping our marginal business.

I put a plan in motion to recoup recent loses of market share and to reduce expenses by twenty percent while increasing income by thirty-five percent through more profitable markets. This was enough to keep me busy, but one more step was necessary to make it work. Consolidation, which had worked so well for the Middle East and Africa, seemed like a good idea here at home.

From 1953 to 1961, I was on the road constantly, travelling from coast to coast as trouble shooter in the massive consolidations we had undertaken. We first merged the St. Louis office with New York. Then San Francisco was merged. Moves and promotions occurred at a dizzying pace. Shell personnel were being moved around like chessmen as we attempted to make the best match between the man and the job. By the early 1960's, the entire U.S. operations were merged into one New York head office.

But in the interim, personnel shifts left some of the other offices looking like Swiss cheese. From 1954 to 1956, while the nation was enjoying peak production and nearly full employment, I acted as

Division Manager for the Cleveland office, taking over for a man who had been kicked upstairs. Then it was back to New York, to give my undivided attention to being General Manager of Head Office Marketing, a job I had continued to fill by remote control in Cleveland.

Moving back to Montclair from Cleveland brought me up sharply to the realization that our lives were changing. Our fourteen-room house seemed cavernous without the kids and their friends. Only Chookey, our fox terrier, scampered through the house. This meant more freedom for us and more free time for Virginia. She spent more time at the piano and soon recaptured her fluid skill, but I found myself listening for Katie's reluctant piano practices and the cacophony of John's Dixieland band. The music room seemed too big and empty now.

Virginia and I filled the empty spaces with our new color TV and with friends. Riding the Lackawanna Railroad into Manhattan every day was a familiar pleasure, like going to one's club. The train was filled with friends and neighbors, everybody riding in the same seats each day, trading news and making social plans. During our daily card games we made weekend golf dates or planned backyard barbecues.

On Friday nights our wives met us at the Montclair station. They were sometimes hard to recognize in the wigs and boots that were in fashion, so several of us developed the safe strategy of seeking out our familiar family car. We would pile into the cars we knew were ours and drive to the house of whichever neighbor was hosting that week's potluck. There was a certain excitement in being driven along familiar streets by an unfamiliar-looking woman, as Elvis Presley crooned "Love Me Tender" on the car radio.

Montclair was a friendly town. Even the morning walk to the station was pleasant and safe. During icy weather, I walked to the station in my golf shoes, carrying street shoes in my coat pocket. At the station, I'd tuck the golf shoes under a bench, certain they would be there when I returned. Those shoes often sat under a bench for days, either because I had flown out of New York on a business trip or because I had driven home with Virginia after meeting her in the city for dinner and the theater.

I suppose our new empty-nest lifestyle indicated approaching middle age, but I didn't have time to think about it. Even the Surgeon General's announcement linking cigarette smoking to cancer failed to impress me as I worried about the cold war, Sputnik, the AFL-CIO merger, and especially the Eisenhower Doctrine. I wasn't at all sure that Eisenhower's gifts to the Middle East were going to solve the problems of the oil business.

While Shell battled the independents, the American majors had the additional problem of dealing with Justice, which was hot on the heels of the FTC report stating that oil was a price-fixing worldwide monopoly. The Justice Department's guns, of course, were aimed at the five American-based companies with global systems of production, transportation, refining and marketing. The very extent of their holdings and commitments limited their former daring: a bold move in one country could have unforeseen repercussions in another.

The independents were taking advantage of this increasing inflexibility and of the hunt by Justice. They depicted the majors as traitors and began angling for more than their Texas oil.

They took exploration risks in the Middle East, offering Arab nations better deals then the majors. In 1954, nine independents joined the majors in forming the Iranian Consortium. In 1956, fifty-one concessions were granted to seventeen companies in Libya.

By 1957, a glut of oil was beginning to force down world prices, and the majors could not control it. The independents were undercutting them and still making big profits.

The Japanese, too, wanted a direct stake in the Middle East and escape from their dependence on the big companies. They offered Saudi and Kuwait fifty-seven percent of the profits and began exporting oil to Japan at a good profit.

Then the independents began competing to sell their Middle East oil to Europe and Japan as they began to be shut out of the U.S. by heavy imports of foreign oil. Several of them, like Hunt and Getty, were now big enough to gain the ear of government. In 1959, import quotas were established, limiting oil from abroad to only thirteen percent of consumption.

The American independents, along with the majors, continued to enjoy the benefits of a strong oil lobby and the advantages of a friendly government. Lyndon B. Johnson, Senate Majority Leader, and Sam Rayburn, Speaker of the House, were Texans and supportive of the oil industry. Wilbur Mills of Arkansas, Chairman of the House Ways and Means Committee, was a sympathetic Congressman. Whatever happened abroad, oil still had the tax advantages at home of the depletion allowance and the intangible drilling allowance. And the foreign tax credit of "Golden Gimmick" fame was whetting a lot of appetites.

In spite of all the pushing and shoving within the industry, everybody was making money. The industry was expanding at about eight percent a year and there seemed to be no end in sight to this growth.

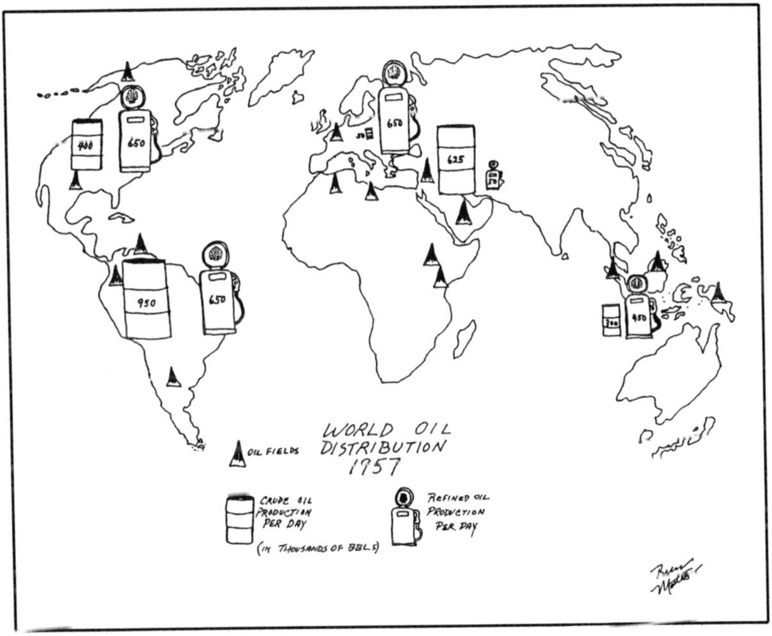

Chapter XXIV

Battling for Crude

In 1958, I was back in London to discuss, among other things, strategies concerning the growing role of the independents in the heating up of the Middle East nationalization movement.

Mayne and Bousfield wore tight smiles as they greeted me in the conference room. It was apparent that the confident expansion plans launched in 1951 were travelling rough seas. Though worldwide consumption had increased as expected, the certainty of available crude was threatened.

"We've done what we set out to do," Mayne stated with a thrust of his chin. "We're leading the market now, everywhere but in South America, and we've built the refineries and marketing organizations needed to double sales in another ten years. But unless we can be certain of enough crude, at low enough prices to be competitive, we may have built ourselves a white elephant." Mayne, a cut-glass Britain, was being unusually candid.

"We counted on stability of supply, but the past few years have shown this to be unrealistic. First, there was the wave of Middle East nationalizations following the Suez War. Then, that Italian oil company began bringing Russian oil into Europe and eroded our position there.

"Now we have trouble within the ranks. Independents, wheedling their way into the Middle East, have pared profits and forced us all to cut posted prices. The Arabs are unhappy. Their oil revenues are set by the posted price, and they may be organizing."

"They're more sophisticated and better educated than they used to be," I observed, "and they're concentrating on gaining expertise in the oil industry. American universities are graduating more and more engineers, geologists, and economists from the Middle East."

"Indeed," Bousfield groaned. "Since 1954, Nasser has been steadily reminding his Arab brothers that it costs only ten cents to extract a barrel of Middle East oil. And they are well acquainted with the antitrust argument in America."

"Yes," I smiled. "The American majors still carry a reputation for market-fixing at home, even though the independents are now setting the benchmark."

Mayne nodded. "That's a boon for U.S. Shell," he said, "but it isn't doing us much good internationally. If Esso's troubles worsen they might go for more reductions in posted prices. The rest of the international companies will have to follow suit, and all hell will break loose with the Arab states."

"Good old Esso," Bousfield chuckled. "It seems that our marketing history has come full cycle, thanks to our perennial competitor. If Rockefeller's old Standard Oil trust hadn't tried to run Royal Dutch out of the Far East sixty years ago, it probably would never have merged with Shell Transport and become an international giant. And now that we have, we're still mightily affected by Rockefeller's offsprings."

"But now," I added, "we're not doing battle as a David to Standard's Goliath. We're an equal power. I spend a lot of time with

Esso execs in New York. They respect us, and they're all too aware of our retaliatory abilities. I think they'll stick to what they can get away with."

"Well, perhaps," Mayne allowed. "But nevertheless we have to protect our sources, and many things threaten that." He paused for a moment, consulting a chart, then continued. "We're taking all the oil we want out of Venezuela and I'm not worried about North America."

"Right," I put in. "U.S. Shell is producing two-thirds of what it needs for its home market, and recent explorations have been very successful. We're adding to our reserves faster than we're depleting them."

"And we've landed the offshore concession of Qatar on the southern Persian Gulf," Bousfield added. "That, plus our fourteen percent in the Iranian Consortium. Our Middle East production adds up to . . . let me see . . ." Bousfield began rifling through the papers in front of him. "335,000 barrels a day from Kuwait, 140,000 from Iraq, 100,000 from Iran and 30,000 from Qatar. That's 605,000 total."

Mayne again consulted his chart, then stated, "Our share of Middle East reserves, that is, oil still underground, is estimated at something like thirty-three billion barrels. That's nearly three-quarters of our total reserves, and it isn't enough. Nasser's nationalization of the Suez Canal and the expulsion of the British makes our access to those reserves uncertain. Then there's that damnable Mattei in Italy. He's upsetting our applecart, talking to the potentates about 'foreign exploitation' and telling them that fifty-fifty is obsolete. He's made a deal with Iran, giving them seventy-five percent!"

"Actually," Bousfield said as he leafed through his papers, "the fifty-fifty principle has not been breached. Mattei's simply made the Iranians a partner in the venture. If and when he strikes oil, the Iranians will have to shell out their half of the development costs to get a partner's share of the net profit, in addition to the usual fifty-fifty income tax. And don't forget, we produce more crude in the Far East than any other company."

"But production there hasn't risen much in recent years and the market is expanding," Mayne pointed out. He got up and paced the length of the conference room.

"There is no way around it," he announced, slapping the conference table with finality. "More and more of our Far East requirements will have to be supplied from the Middle East. We've got enough crude for today and it is well distributed. Right now we're in a nicely balanced position. But can we continue to maintain twelve years of proven reserves ahead of consumption, and to add more to reserves each year than we take from them?"

"At the rate we're spending, I don't see why not." Bousfield shoved a paper across the conference table to Mayne. "Over the past five years we've spent $756 million on exploration and oil rights."

"But most of those explorations are in partnerships with other companies, and in today's changing climate I'm not sure we can count on them."

"True," Bousfield allowed, "but our fields in Algeria and Nigeria have finally struck important oil. In a few years, they could exceed production in Borneo and Indonesia."

They were back onto something I knew well. "Before we can be sure of that," I told them, "we'll have to build a pipeline to the terminal and test the wells under continuous use for at least a month."

"And we'll have to increase our exploration and drilling," Mayne insisted. "Any oil company that aims to double its sales simply cannot continue to take chances on giving up good crude. Anywhere. Our tanker fleets can move it wherever we want it, but we've got to be assured of having it."

Mayne pushed a button on the arm of his chair. A moment later, Geoffrey entered the room.

"Geoffrey, take a memo to the other directors."

On the plane home, I read a Belgian analysis of the oil situation. It was a dizzying circular story of oil as the pawn in an escalating tug of war between shifting Arab allies and disunified Western powers.

Arab hatred for Israel was being fomented by Nasser. He was calling for Arab solidarity in destroying Israel. Nasser had strong-armed Saudi into stopping the sale of oil to the French and British after their unsuccessful resistance to his nationalization of the Suez Canal. Israel, who had joined the French and British attack, had been denied use of the Canal.

King Saud, son and successor of Abdul Aziz, was furious about Nasser's move, which resulted in a forty percent drop in Saudi oil revenues at a time when the king was heavily in debt. Saud was jealous of Nasser's popularity throughout the Middle East but he had no choice but to support him. When Iraq had been slow to back him, Nasser had blown up the their pipeline and Saud didn't want to risk the same thing for his own pipelines.

This is where the analysis brought Western disunity into the picture. The U.S. had urged Britain to evacuate the Canal after its nationalization, and when Britain complied, this had destroyed British influence in the Middle East. Then, concerned about its oil interests, which were no longer stabilized by a British and French presence in the area, the U.S. had given King Saud a $180 million increase in aid and backed him as a moderate alternative to Nasser. Within the U.S., this aid was not popular and Congress didn't like it.

"Here we are," Oregon Senator Wayne Morse was quoted, "pouring by way of gifts to that completely totalitarian state, Saudi Arabia, millions of dollars of the taxpayers' money to maintain the military forces of a dictatorship. We ought to have our heads examined!"

But the deal was made, and Saud was now openly opposing Nasser. Whether this was for his own purposes or because of allegiance to the U.S. remained to be seen. What was certain was that the Suez crisis had resulted in unstable oil prices.

The Arab states were on an economic seesaw controlled by the oil companies, who were themselves jostled around by Western and Middle East governments. The Arabs didn't like this new fluctuation in their oil incomes and were listening to new voices. Blaming the established oil companies for their new troubles, Arab nations began entertaining new offers from independents around the world.

Not only were the independents threatening the majors' markets, they were doing their best to grab our oil fields for themselves. Mayne was right to divert marketing funds into further oil field development. I wondered what effect this would have on my U.S. Marketing budget.

I closed my eyes and let the hum of the plane's engines soothe me. Just before dropping off to sleep, a wisp of old conversation went through my mind. I heard Bousfield in 1950, in my living room in

Montclair: "A lot more energy will be put into exploration and production," he was saying. "The companies will be preoccupied with crude." I drifted off, wondering if he were a wizard or simply the most astute mortal man I had ever known.

In 1959, the posted price of oil was cut by eighteen cents a barrel. In 1960, Esso cut its Middle East posted prices by ten cents a barrel. The other majors followed, and Venezuela's long courtship of the Arab nations kicked in. The Venezuelans met with Iran, Saudi, Iraq and Kuwait—and OPEC, a cartel to confront a cartel, was formed.

In their first resolution, the exporting countries made clear that their paramount enemy was the oil companies. The resolution stated:

> *. . . that members can no longer remain indifferent to the attitude heretofore adopted by the oil companies in affecting price modifications; that members shall demand that oil companies maintain their prices steady and free from all unnecessary fluctuation; that members shall endeavor, by all means available to them, to restore present prices to the levels prevailing before the reductions . . . .*

The statements of the newly formed Organization of Petroleum Exporting Countries made clear the common problems of the oil exporters. They all depended on oil income to finance their development and balance their budgets; oil was a depleting asset which must be replaced by other assets; fluctuations in the price of oil dislocated their economies as well as the economies of all consuming nations.

Mayne had foreseen this and I had been wrong about Esso. The company's unilateral decision to cut prices, without consultation with the oil producers, had compelled the producers to react in kind. This cut had precipitated the producers militant response and had changed the whole world view of all the major oil companies.

The facts from the Federal Trade Commission's report in 1952 were repeated in OPEC speeches. Like the anticolonialists in the British Empire, they had taken their weapons from their masters. They could count on support from the consuming countries in putting forward their cartel. When the industry was attacked, it was without friends.

OPEC succeeded in preventing further reductions in the posted price. The oil companies tried to get individual countries to give tax rebates in return for extra production. But the OPEC countries stood firm; but they did not restore the posted prices to the previous levels, and its members could not agree on how to fix prices or restrict production. Each member pushed for a bigger share of the market, and because of this disunity the oil companies gained back some of their control of production.

Chapter XXV

World in Transition

Chastened by my misreading of Esso, I put my passport away and returned to the problems of consolidating U.S. marketing. My suitcases, however, remained in use.

There was the usual jumping around as troubleshooter and sometimes having to leap into a vacancy. Filling in for the regular VP who was going abroad, I took over as Vice President for the Central Marketing Region in Chicago in 1961. From 1962 until 1965, I was in San Francisco, as Vice President of Western Marketing to finalize the merger and the transfer of personnel. These tasks were complicated by high national unemployment and the increased income taxes that had accompanied Kennedy's Presidency, and by the embargo on exports to Cuba.

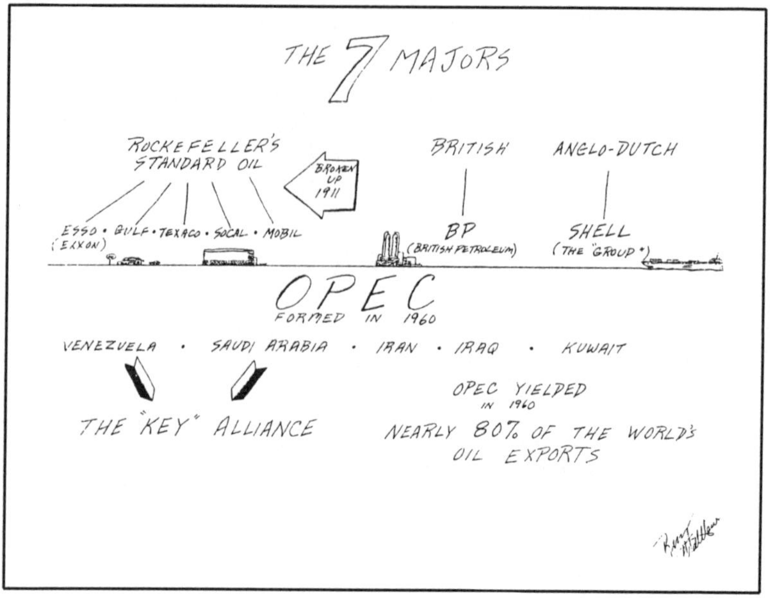

During these years, Katie had spent some time in France, which signaled an end to adolescence. She had graduated from Northwestern in 1956 and was working for Time/Life in Rockefeller Center. When I was in New York, I often met her for dinner at the club, where she tried repeatedly and unsuccessfully to teach me to dance the "twist." Suddenly I realized she was no longer a kid when acquaintances mistook her for my girlfriend.

John had completed his tour of duty with the Marines, earned his engineering degree from Purdue, and married. He was living in San Francisco so we saw a lot of him, especially on weekends at the mountain cabin Virginia and I owned in Boulder Creek.

The 1950's had flown by, with bigger international and domestic responsibilities for me and in the States during this period of headlong expansion. The formation of OPEC in 1960 hit most of the majors like a punch in the gut, but had minimal effect on Shell. We had guessed right and now were on top in sales transportation. It made little difference whether we bought our crude from Aramco and British Petroleum or from the oil-producing countries themselves, so long as there was crude to buy.

The Suez invasion in 1956 had brought price hikes, and we were still dealing with price wars created by the independents, but in general we were sitting pretty. The world was experiencing an unprecedented oil glut. With Shell's concentration on transportation and marketing, we were on top of the heap.

OPEC had succeeded in preventing further reductions in posted prices despite the intensified competition, but was unsuccessful at establishing quotas on its oil production. The big producers, like Saudi, did not want to be limited. The oil companies were trying to hold back production in the Middle East, keeping such potential in reserve, while the OPEC countries were trying to push it for increased revenues.

Since all the majors were interlocked in joint ventures around the world, their goal was more manageable. Their strategy was to divide and conquer the oil potentates, and they were again being accused of restraint of trade.

The attorney who represented the antitrust interests of the seven majors went to see President John F. Kennedy. He was worried about a Middle East confrontation with the Russians, and the attorney warned that the oil companies might need authority for collective bargaining if OPEC succeeded in forming a price cartel. To secure oil supplies and for reasons of foreign policy, the antitrust laws were again waived.

Since World War II, the seven majors had managed a remarkably constant increase in output, averaging nearly ten percent a year. But unlike the pre-war cartel days, they had not been able to keep prices up to artificial levels. By the late 1960's, it was hard to see how prices could fall much lower.

Meanwhile, growing tension between the Israelis and Arabs was short-circuiting the opposing elements of American foreign policy. In 1967, Israel single-handedly defeated Egypt, Jordan and Syria, capturing the Sinai and Gaza Strip from Egypt, the Golan Heights from Syria, and the West Bank and Jerusalem from Jordan. The Arab humiliation was even greater than it had been 1948.

Israel had captured a territory three times larger than its original size. Israeli intellectuals stated that Israel had "reached its natural borders, and liberated the historic homeland of the Jewish people, removing the Syrian threat and turning the Suez Canal into a genuine international waterway."

The Arab world rejected talks of peace in exchange for the Sinai and the Golan Heights. "What has been taken by force will be retrieved by force," affirmed Egyptian leader Gamel Abdul Nasser.

Arabs longing for retribution were now red-hot with anger at their huge defeat and by Israel's determination to keep the West Bank and Jerusalem. The land that was considered the cradle of civilization by the Jewish nation could not be negotiated or abandoned. Jerusalem, the site of the two temples and King Solomon's capital, was the eternal symbol of the Jews. A few weeks after the six-day war, Israel annexed Jerusalem.

The Arabs, believing Britain and the U.S. were supporting Israel, agreed to shut down their oil fields in an attempt to boycott the Western imperialists, but Iran and Venezuela would not join the embargo. A month later the oil fields were back to normal exports.

I was in Florida, gazing out at the Gulf of Mexico, trying to figure out everything that was going on as I juggled phone calls and appointments with building contractors who were working on my new home. It was 1967, and the Beatles had started mixing music from India into songs that didn't make sense—or did they?

I was preparing to retire and had time to think about how modern communications technology was causing everything to change so fast. My whole career had been one of meetings, trips, collaborations, and completing projects. It had all been face-to-face, it was personal, it seemed so real compared to what was happening now. I thought about the ideas of Marshall McLuhan, that the medium (of telecommunications) is the message itself. I still couldn't quite figure out what McLuhan had meant by that, but I did know that it probably wasn't going to be as easy, from now on, for a small group of Western oil men, political leaders, to make history without outside reference. Was it just that the Arabs were beginning to grasp the magnitude of their own power, or had we given them a gift—telecommunications—that no man, nation or power bloc could control?

Chapter XXVI

Culminations and Losses

I have neglected to tell you how I had ended up looking out over the Gulf of Mexico. Back in 1958, the purpose of my trip to London had been to report on conditions in their Puerto Rican company.

I had been keeping a cursory eye on Puerto Rico for the Group since 1952 and reported that Shell needed to update its marketing and operations there. Now that Castro had nationalized Shell's Cuba holdings, Puerto Rico Shell would have to take over for Shell Cuba. The Group felt that someone from U.S. Shell could best do the job.

There were labor troubles in Puerto Rico. London needed someone to handle them, someone nearby who could work with the people there. I volunteered, was made a director of the company, and began making several trips a year from New York. When the

Group reorganized in 1960, Puerto Rico reported to U.S. Shell, which was responsible for many smaller subsidiaries throughout the Caribbean. Such changes brought to me an additional title—Chairman of the Board.

By 1965, two years before my retirement, I was working mostly as a consultant, doing some travelling but mostly winding up, preparing to turn things over to my successor. It had been a long time since I had spent such long periods of time in New York. I was starting to feel bored, even though I was ready to slow down.

I had begun noticing the changes taking place in the world around me. I was charmed and disturbed by mini-skirts. The bell-bottom trousers the men were wearing reminded me of World War II, but the long-haired youths I saw everywhere, the drugs, and the urban chaos that were spreading across the country, seemed to reflect profound changes in the American landscape. I felt I could hardly recognize my own country. Walking down Fifth Avenue, I felt like a tourist in an alien and unpleasant land.

Shell has always been good to its people. Their company policies, generous and very considerate for people heading into retirement, helped me get through the feelings of awkwardness and uncertainty of my impending separation.

Over lunch one day, my boss found out that Virginia and I were building a retirement home in Florida. He was quick with an idea: "Got a deal for you. You know that pipeline we're building in Puerto Rico? Nobody knows pipelines like you do, Jack. You've already done a good job there smoothing government feathers with the labor problems. We need someone to reside in Puerto Rico for the stockholders' meetings. Why not you? It's only a short hop to Florida. It'll be easy work, in a pleasant environment, and you'll be able to fly up to Naples on weekends."

I talked it over with Virginia. A few months later we sold our Montclair house and our cabin in Boulder Creek, shipped our cars to Puerto Rico, and moved into a San Juan penthouse that Shell leased for us.

A telegram from Bousfield awaited our arrival. It said: "CONGRATULATIONS ON CHAIRMANSHIP. BETTER THAN MERE BOARD MEMBERSHIP. AM EATING CROW."

Puerto Rico was easy duty. The government backed me in handling the labor problems and the pipeline was a cinch. I spent a lot of time entertaining in this international travel center, with Shell providing membership in six Puerto Rican clubs. With company guests and many friends visiting from around the world, Virginia and I played a lot of golf, attended many parties, and frequently vacationed in St. Croix and St. Thomas. We spent our weekends in Naples, working with our architect and contractor on our house.

The unhurried Caribbean climate helped me to relax to the pace of "island time." It was hard to get excited about the oil majors' declining profits, brought on by the independents and OPEC. I figured this was only a temporary glitch. The majors were still the big boys on the block. Soon, able younger men in the companies would be getting things back on track.

I was quite content supervising the Puerto Rico pipeline construction and the tanker runs to the nearby islands. Meanwhile, I continued to monitor the construction of our Naples home. I would be ending my career in November 1967. Even the six-day war in the Middle East failed to dampen my enthusiasm of preparing for retirement.

We moved to Naples on schedule, while the exterior of the house was still unfinished and landscaping to be done. These final touches, along with installing a swimming pool, kept me busy for the first few months. Virginia, meanwhile, became a Pink Lady at the Naples hospital, joined a women's club and golf club, and talked me into joining the yacht club.

Once the house and grounds were finished I didn't know what to do with myself. My adrenaline was still pumping but there was nothing to do with it. I watched boats cruise up and down the channel behind our newly landscaped yard. It made me grumpy. I yelled at Virginia a few times over nothing until she finally said, "Go find something to do before you drive us both crazy."

She was right. I had watched too many retired friends drink themselves into their graves, victims of "Sudden Sedentary Syndrome." I remembered a saying of my mother's, "The time you enjoy wasting is not a waste of time." It was for me time to reevaluate.

I began playing golf. One day, when asked to join a foursome for a tournament, I met Leston Faneuf, retired Chairman of the Board of Bell Aircraft. Leston, an athletic man with a romantic bushy mustache, turned that foursome into the most interesting golf game of my life.

As we progressed from tee to tee, Leston regaled us with stories of salmon fishing in all the famous rivers of Europe and Canada. Afterwards, we returned to the clubhouse and urged him on. We were rewarded with tales of big game and movie safaris in Africa.

Virginia had often played golf with his wife, Millie, who had accompanied Leston on his fishing trips and safaris. When I told her how much I had enjoyed Leston's stories, Virginia invited them to dinner. It was then we learned that Millie was an excellent fly fisherwoman and was well respected for her marksmanship by the professional hunters in the African bush. In fact, Leston bragged, it was Millie who handled their yacht, Emelef, when the captain was not aboard. We began dining with them regularly and the four of us soon became close friends.

Leston and I, along with eight neighbors all needing something to work on, formed the Port Royal Property Owners Association. Working with the landscaping people, the city and the police took some of the boredom out of my enforced leisure. I had a project again, and dove into the work of being the Association's contact with local government. Then I became a volunteer for the Naples hospital. I now had enough to keep me pleasantly busy, without the deadlines and strains of the executive life.

There was time every day to watch the magnificent Florida sunsets, to read, to simply sit and think. Retirement was beginning to take on a rosy glow of well-being as I discovered the natural rhythms of life.

This was a wonderful time for Virginia and me. For the first time ever, we had unlimited time together, and we fell in love in a new way, deep and sweet, filled with the trust we had built during our forty-year partnership. Virginia was relieved to be finally settled in a permanent home. I was content as secretary and head of building maintenance for the Association and as head of the hospital's Men's Auxiliary. We made a lot of good friends so we felt no urge to travel, other than to visit our children and grandchildren. Even Millie and Leston's invitation to spend August with them in a castle they had rented in Austria failed to tempt us from our plans to visit John and his family in California. Life seemed perfect.

One morning in California, Virginia woke in great pain. She thought it was her gall bladder so we called our doctor in Naples who referred us to a California physician. He gave her medication to ease the pain and we flew home for three more rounds of tests and diagnosis. They each confirmed the California doctor's prognosis. Virginia had cancer of the kidneys and pancreas and she went on cobalt treatments. Although she looked healthy enough, her life became a seesaw of good days and bad days.

Our fiftieth anniversary was too far away so we decided to throw a huge party on our fortieth in 1971. We hosted a large black-tie affair at the Port Royal Club. Virginia was excited about the party, and despite feeling tired the day it was held, she insisted on attending and managed to be a smiling and gracious hostess for the event. By 10:30 P.M., she was exhausted so we excused ourselves and went home, leaving our friends to toast our forty years of happiness.

She stayed in bed for a week after that, and her doctor recommended surgery at her next appointment. The surgery confirmed that the cancer was terminal.

Virginia was very brave through her rounds of hospitalization and, when at home, had round-the-clock nursing care. Because of my hospital work, I was able to spend as much time with her in the hospital as when she was at home. By then, I was a trustee of the hospital and Virginia was given one of the best rooms. She wanted for nothing.

I saw a lot of Millie Faneuf, whose husband, Leston was also dying. Leston's room was on the same floor as Virginia's, Millie and I often shared nightly vigils in the family waiting room, trying to keep up each other's spirits.

When she was home, Virginia and I spent evenings together. She wanted a million things done before she died: notes sent to cousins and old friends, arrangements made for Katie, John, other family members and friends to be given her most precious possessions. She was handing our social responsibilities over to me. I was unpracticed and awkward in dealing with them.

Throughout our marriage Virginia had always managed our social relationships and she was still trying to shoulder them. She wouldn't allow the kids to visit during her bad spells. When they did come, she put on a good appearance, never letting them see the painful side of her illness.

She insisted I keep up with my civic activities. I did so to please her. A neighbor, Naomi Williams, spent a lot of time with her when I was out. I felt numb but went through the motions of staying busy until Virginia's final two weeks, when I lived at the hospital.

Katie and her husband Dave came to the funeral, just before Christmas in 1973. John was there, too, and insisted I go back to San Francisco to spend Christmas with him and his family. His daughter Amy was five and Annie was only eighteen months old. Being with them took the grimness out of Christmas for me, but going back to our home in Naples was the most difficult thing I had ever done.

Chapter XXVII

Starting Over

Back in our dream house in Naples, the aloneness hit me. I wandered around the house, aimlessly picking up Virginia's hairbrush or her knitting. Sometimes I would find an hour had passed before I put them down again. Sometimes I just stood looking into her closet or at her perfume bottles on the dresser. Finally, some neighbors rescued me by packing up Virginia's personal things and sending them off to the hospital's White Elephant resale shop.

My neighbors' actions helped, but I remained stuck in the habits and dependencies of forty years. The cleaning woman still came twice a week, but I was unable to give the instructions she needed. How often should books be dusted? When should the carpeting and upholstery be cleaned? She managed to keep the place picked up, but it began to take on a neglected air.

For the first time, I began to realize how much went into running a home, and when friends and relatives began to complain about not hearing from me, I understood the responsibilities of social relationships that Virginia had handled so smoothly for us.

I missed all she had done for us and her cheerful companionship. I became filled with a deep regret that so much of our life together had been taken away by my job.

The hospital and Association work kept me from going crazy. Still nothing eased my nighttime despondency until Millie returned to town. Leston had died three months before Virginia, and Millie had been with her family in New York.

When I heard she was back, it seemed like the most natural thing in the world to call her. Who understood Millie's loss better than I? Who understood her better now that Leston was gone?

During our shared hospital vigils, neither of us had been capable of keeping up social fronts. We talked openly to each other about our sorrows and our fears. We had been friends for years, but this experience had brought us to something closer than friendship. In our pain, neither of us tried to put a name on it. We had simply and gratefully accepted the comfort we had given each other. Each of us was losing a friend and mate. We each felt profound true grief at the other's loss.

Our first date was on a Sunday morning. We went to church together. Word got around and friends began inviting us to gatherings as a couple.

Our new relationship went easily, fitting without effort into the pattern of our former lives. We knew all the same people and were involved in the same causes. Millie took over the supervision of my maid. Little by little, I came to depend on Millie in subtle ways. I was perfectly content until one afternoon when my neighbor, Betty Butler, stopped by for a chat.

I was starting to get the hang of bachelorhood and remembered to offer Betty a drink. Settling in the Florida room that looked out over the swimming pool and boat channel beyond, we exchanged news of this and that. Betty, a friend of Millie's best friend, said that Millie had a big decision to make.

It seemed that a good friend of both Millie and Leston, a general in the South African army, had lost his wife about the same time Leston had died. The general and his wife had often joined Millie and Leston in Europe, and had entertained them royally in both Cape Town and Pretoria.

This general had asked Millie to come to South Africa and be his wife. Millie knew that a good life awaited her there, Betty continued, and was considering his proposal.

I lost no time in proposing. On February 21, 1974, we were married in Trinity-by-the-Cove Episcopal Church, the same church we had attended on our first date.

My daughter, Katie, was matron of honor and Millie's brother, James, was our best man. It was a quiet family wedding, but many of our friends were waiting outside the church to greet us after the ceremony. As we stopped to receive their congratulations and I heard James quip, "My sister is young enough to marry again, but too old to wait."

After hugging Millie, Betty Butler said to me, "Do you realize, Jack, that in marrying Millie you have also married her maid, her cat and 1,500 books?"

Even our contractor was there, and Millie and I mentioned in passing that we wanted him to build a house for us.

"Why?" he asked. "You each own one of the best built houses in Port Royal. Sell one and live in the other."

So Millie sold her Blue Shutters and moved her maid, her many books, and Princess, the cat, into the house that Jack built.

We solved the book problem by turning one of the three bedrooms into a library. Princess was another matter. For months, no matter how many times we brought her to her new home, she would turn around and return to Blue Shutters. Millie's expertise with animals had been with show dogs, horses, and Jersey cattle. She had no idea what to do about that cat's disposition to return her to her former home. Katie, experienced with cats, solved this for us on one of her visits.

"Forget about her," Katie told Millie. "Leave her there and wait. She'll come back to her new home." And she did.

We kept both our maids, combined our furniture, and added a few new pieces. Soon we had a home and life of our own. The three of us. I became very fond of Princess who, with her pink leather rhinestone collar and leash, was a real showstopper. As the three of us took our nightly walks, cars would stop to watch this cat on a leash.

We had blended our lives easily and quietly, with only one little bit of commotion. This came when we took Leston's fabulous gun collection to the trustees of his estate, for shipment to Leston's son-in-law. As Millie and I, escorted by a bank guard, carried the guns into the trust department, a bank employee fainted thinking that these elephant guns and big game rifles were part of a hold-up.

We combined our club membership in Naples and spent our summers at the Wilmurt Club in the Adirondacks. Millie and Leston had been members. Often we visited Millie's brother, James, who had a summer place on Lake Ontario.

Our summer life on the mountaintop in the Adirondacks was very active: filled with boating, long walks through the woods, afternoon bridge games and social affairs in the evenings, either at our house or the house of another member. Special occasions were the big parties on the Fourth of July and Labor Day, and our yearly visits to Hamilton, New York, to visit good friends who were former Montclair neighbors.

Our life in Naples did not change much from the days when Virginia and Leston had been with us, except in two respects. I found myself attending the ten fashion shows each year in which Millie modeled for charity. We entertained more, often hosting dinner parties for twelve. Once a year we threw a black-tie party for eighty guests, at either the Naples Yacht Club or the Port Royal Club.

But nothing lasts forever. After fourteen years of this pleasant life, with summer sojourns to the Adirondacks mixed with visits to John and his family in California and to Katie and her family in Illinois, we settled into year-round life in Naples. I now had a new and final project: controlling the Parkinson's Disease that had crept up on me, hidden behind my osteoarthritis for two years before it was confirmed.

They caught my Parkinson's Disease early enough to treat it with medication, exercise and a prescribed diet. Parkinsonism is described by some doctors as a state of brain Dopamine depletion. I experience loss of balance and slowness of movement, but as long as I had something to hold on to, I was fine. I could still dress myself, drive, and lead an active social life.

Millie, a glamorous and exciting companion all these years, showed the additional talent of being a patient and supportive friend and caregiver throughout. Formerly active in eight charities and serving on the board of directors of four of them, Millie limited her outside activities to volunteer work at the Naples Community Hospital and to some activities at Trinity-by-the-Cove Episcopal Church. The rest of her considerable talents are now centered in our home. Long a favorite with my family, she arranged frequent visits from my kids and grandkids.

Dancing and golfing were out, replaced by mile-long walks and a prescribed daily exercise program. But there were still games of bridge with friends, lunches at the club, evenings at the Philharmonic, and time left over to keep up with the world and to write this book. She devoted every bit of her energy to making sure I would finish this manuscript. Here I am, I know that it is done and for that I want to thank Millie.

Postlogue

Jack Hall finished his autobiography in 1991 and continued leading an active life until October of 1992, when he was hospitalized. He rallied and two months later was able to return home under the care of round-the-clock nurses.

His home stay was punctuated by his perennial good humor and a continued keen interest in the world around him.

In mid-March he returned to the hospital and one week later, on March 22, 1993, Jack quietly succumbed to pneumonia.

His brave and generous spirit is missed by all who knew him

and especially by Millie, who was his constant companion and comfort to the end. Here is her final poem to him, written an hour after his death:

Goodbye, My Love

I said Goodbye,
stroked your head,
patted your hand
and said I love you.

You did not hear or know me.

We had nineteen years
of happiness:
no cross words
or arguments—
difficult problems
solved intelligently.

Someday I'll join you.
Goodbye, my love.

Your wife,

Millie

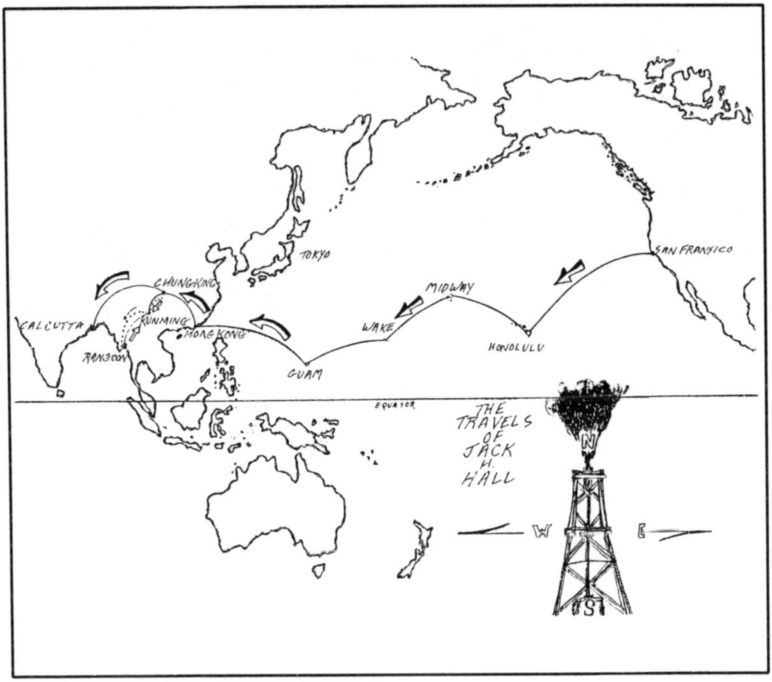
TOKYO
SAN FRANSICO
CHUNGKING
MIDWAY
CALCUTTA
KUNMING
HONG KONG
WAKE
HONOLULU
RANGOON
GUAM
EQUATOR
THE TRAVELS OF JACK H. HALL
N
W
E
S

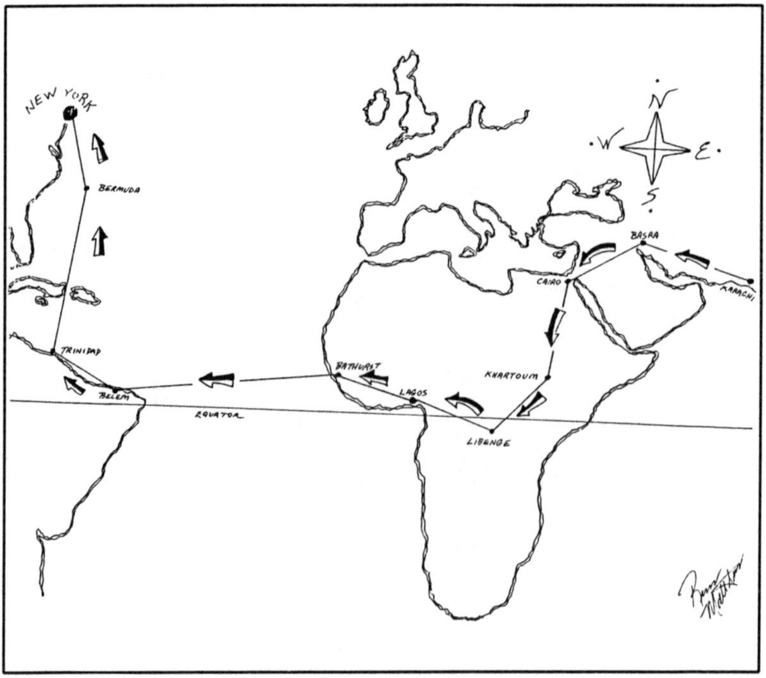
NEW YORK
BERMUDA
TRINIDAD
BELEM
EQUATOR
BATHURST
LAGOS
LIBENGE
KHARTOUM
CAIRO
BASRA
KARACHI
N
W
E
S